Historical Essays on Radioactivity

Scientiarum Historia et Theoria Studia, volume 2

Roberto de Andrade Martins

Historical Essays on Radioactivity

Extrema
Quamcumque Editum
2021

Print edition – ISBN: 978-65-996890-2-4
E-book edition – ISBN: 978-65-996890-5-5

Summary

FOREWORD

This volume comprises four historical papers concerning the early researches on radioactivity. Three of them were written in 1995-1996 but have never been published, for a variety of reasons that will not be described here. Due to my advanced age and other current circumstances, it would not be wise to postpone the publication of those works even more. Hence they are released here, without much alteration. The fourth one was produced in 2005.

The history of radioactivity – especially its discovery and early investigations – was an old concern of mine.[1] It was quite difficult, however, to develop a historical research on this theme using the resources that were available in Brazil, before the accessibility of digital libraries. During the period 1995-1996, I was a visiting scholar of the Department of History and Philosophy of Science, University of Cambridge, and visiting fellow of Wolfson College, and I benefitted from this opportunity to advance my studies on this subject, besides investigating the history of experimental gravitation in the early twentieth century.[2] In that period I produced four articles, but only one of them was published.[3] No attempt was made to update them or to complement the research they describe. They are reproduced here, with slight changes, for the first time.

---

[1] My first published paper on this subject was this: MARTINS, Roberto de Andrade. Como Becquerel não descobriu a radioatividade. *Caderno Catarinense de Ensino de Física* **7**: 27-45, 1990.

[2] See the first volume of this series: *Studies in History and Philosophy of Science I.*

[3] MARTINS, Roberto de Andrade. Becquerel and the choice of uranium compounds. *Archive for History of Exact Sciences* **51** (1): 67-81, 1997.

(1) A pool of radiations: Becquerel and Poincaré's conjecture

This paper presents the intricate context of the discovery of radioactivity. A short account of its results was published in Portuguese, a few years later.[4] A fuller version appeared in the second chapter of my book on the history of radioactivity (in Portuguese).[5] Poincaré's conjecture was linked to the discovery of X-rays. I have published a few papers on the early history of X-rays, in Portuguese, which supplement this article.[6]

(2) Did Niepce de Saint-Victor discover radioactivity?

The analysis of the concept of scientific discovery contained in this paper was published, in Spanish, a few years later.[7] The main substance of the article appeared in Portugese, as chapter 7 of my book on the history of radioactivity.

---

[4] MARTINS, Roberto de Andrade. Hipóteses e interpretação experimental: a conjetura de Poincaré e a descoberta da hiperfosforescência por Becquerel e Thompson. *Ciência & Educação* **10** (3): 501-516, 2004.

[5] MARTINS, Roberto de Andrade. *Becquerel e a descoberta da radioatividade: uma análise crítica*. Campina Grande: Editora da UEPB; São Paulo: Livraria da Física, 2012.

[6] MARTINS, Roberto de Andrade. Investigando o invisível: as pesquisas sobre raios X logo após a sua descoberta por Röntgen. *Revista da Sociedade Brasileira de História da Ciência* (17): 81-102, 1997; MARTINS, Roberto de Andrade. A descoberta dos raios X: o primeiro comunicado de Röntgen. *Revista Brasileira de Ensino de Física* **20** (4): 373-91, 1998; MARTINS, Roberto de Andrade. Jevons e o papel da analogia na arte da descoberta experimental: o caso da descoberta dos raios X e sua investigação pré-teórica. *Episteme. Filosofia e História das Ciências em Revista* **3** (6): 222-49, 1998.

[7] MARTINS, Roberto de Andrade. ¿Que es el descubrimiento científico de un nuevo fenómeno? Vol. 5, pp. 281-288, in: SOTA, Eduardo; URTUBEY, Luis (eds.). *Epistemología e Historia de la Ciencia. Selección de Trabajos de las IX Jornadas*. Córdoba: Universidad Nacional de Córdoba, 1999.

(3) Becquerel's experimental mistakes

I intended to present a version of this paper, with the title "A critical analysis of Becquerel's early experiments" at a conference held in Paris, in 1997.[8] However, I could not attend the conference. A short report was presented on my behalf by professor Erwin Hiebert, of Cambridge. A quite small version was published in Spanish.[9] Part of its results are contained in a book chapter published in 1911.[10] Most of its content appeared in Portuguese, in chapters 4 and 5 of my book on the history of radioactivity. The involvement of Henri Becquerel and Jean Becquerel with Blondlot's N-rays was dealt in a more detailed way in a book published in Portuguese.[11]

(4) The guiding hypothesis of the Curies' radioactivity research: secondary X-rays and the Sagnac connection

This paper was written 1n 2005, and a short version was presented at the British Society for the History of Science Annual Conference University of Leeds, 15-17 July 2005. A small portion of its results had already been published in Portuguese, in a paper on Marie Curie's early researches on

---

[8] International Conference on Radioactivity: History and Culture (1896-1930s). Held at the Institut Curie, Paris, 7 to 9 July 1997.

[9] MARTINS, Roberto de Andrade. Los errores experimentales de Henri Becquerel. Vol. 6, pp. 267-274, in: GARCÍA, Pío; MENNA, Sergio H.; RODRÍGUEZ, Víctor (eds.). *Epistemología e Historia de la Ciencia. Selección de Trabajos de las X Jornadas*. Córdoba: Universidad Nacional de Córdoba, 2000.

[10] MARTINS, Roberto de Andrade. Henri Becquerel and radioactivity: a critical revision. Pp. 107-117, in: KRAUSE, Décio; VIDEIRA, Antonio Augusto Passos (orgs.). *Brazilian Studies in Philosophy and History of Science*. New York: Springer, 2011.

[11] MARTINS, Roberto de Andrade. *Os "raios N" de René Blondlot: uma anomalia na história da física*. Rio de Janeiro: Booklink; São Paulo, FAPESP; Campinas, GHTC, 2007.

radioactivity.[12] Chapter 6 of my book on the history of radioactivity, published in 2012, in Portuguese, contains a large part of this paper. The full article was later published as a book chapter.[13] It is reproduced here, with slight changes, with due permission of College Publications.

About the author

Roberto de Andrade Martins is a Brazilian scholar. His research subjects are: foundations of physics (especially relativity theory), history of science (especially history of physics, mathematics, chemistry, astronomy and biology), and philosophy of science. He has published over 200 works on those subjects, and he has supervised over 30 dissertations and theses on his research subjects.

He obtained his first degree in Physics at the University of São Paulo (USP, 1972), and a PhD in Logic and Philosophy of Science at the State University of Campinas (UNICAMP, 1987). He was a member of the faculties of the State University of Londrina (UEL), of the Federal University of Paraná (UFPR) and of the State University of Campinas (UNICAMP), where he worked throughout most of his academic career. After retiring from UNICAMP, in 2010, he was a visiting professor of the State University of Paraíba (UEPB), of the University of São Paulo at São Carlos (USP) and of the Federal University of São Carlos (UFSCar). Since 2016 he is a collaborator of the Federal University of São Paulo (UNIFESP).

He was a research fellow of the Brazilian National Council for Scientific and Technological Development (CNPq) for more

---

[12] MARTINS, Roberto de Andrade. As primeiras investigações de Marie Curie sobre elementos radioativos. *Revista da Sociedade Brasileira de História da Ciência* [series 2] **1** (1): 29-41, 2003.

[13] MARTINS, Roberto de Andrade. The guiding hypothesis of the Curies' radioactivity research: secondary X-rays and the Sagnac connection. Pp. 45-65, in: MARTINS, Roberto de Andrade; BOIDO, Guillermo; RODRÍGUEZ, Víctor (eds.). *History and Philosophy of Physics in the South Cone*. London: College Publications, 2013.

than 40 years. He was president of the Brazilian Society for History of Science (SBHC) and of the South Cone Association for Philosophy and History of Science (AFHIC).

A POOL OF RADIATIONS: BECQUEREL AND POINCARÉ'S CONJECTURE

Roberto de Andrade Martins

Abstract: A few weeks after Wilhelm Conrad Röntgen communicated the discovery of X-rays, Henri Poincaré suggested that the emission of those rays could be associated to luminescence phenomena. Following Poincaré's conjecture, many researchers investigated the emission of penetrating radiation by phosphors and other substances. The "discovery" of several new effect was reported – such as the emission of radiation by sugar and glow worms. The paper describes Henri Becquerel's and Silvanus Thompson's study of the radiation of uranium compounds in this context. Both adopted the natural hypothesis that the observed phenomenon was due to a special phosphorescence that violated Stokes' law. Becquerel, Thompson and contemporaneous scientists were unable to distinguish what we now call radioactivity from other spurious reported phenomena. Those researches were made in a highly speculative and uncritical period, when elementary experimental precautions were overlooked.
Keywords: radioactivity; penetrating radiation; X-rays; history of physics; Becquerel, Henri; Thompson, Silvanus

> *My intention in publishing this Budget [...] is to enable those who have been puzzled by one or two discoverers to see how they look in the lump.*
> (Morgan, 1872, p. 5)

MARTINS, Roberto de Andrade. *Historical Essays on Radioactivity*. Extrema: Quamcumque Editum, 2021.

1. INTRODUCTION

The discovery of radioactivity happened in the exciting years following Wilhelm Conrad Röntgen's announcement of X-rays (Röntgen, 1895; Satson, 1945).[1]

It is well known that in the beginning of January 1896 Röntgen sent pre-prints of his first article to the main scientific leaders of that time. A few weeks later, his work was discussed and reproduced all over the world (Jauncay, 1945; Sarton, 1937). Over a thousand papers on X-rays were published during 1896.

In his first paper, Röntgen had already established many physical properties of the X-rays, but for several years the nature of this radiation remained unknown[2]. It was also unknown how they were produced by the electric discharge in the low-pressure Crookes tubes. It was the discussion about the origin of X-rays that led (among other things) to Becquerel's researches on the radiation of uranium compounds.

The aim of this paper is to describe Becquerel's discovery in its proper scientific context, such as contemporaneous researchers regarded it: as one of several instances of emission of penetrating radiation by luminescent bodies.[3] In this perspective, one is able to see why Becquerel's rays were not at first regarded as a revolutionary, outstanding discovery.

---

[1] The standard scientific biography of Röntgen was written by Otto Glasser (1933). In this book one can find a translation of Röntgen's relevant papers, together with a discussion of the context of the discovery and early consequences.

[2] At first, there were four main hypotheses concerning the nature of X-rays: (a) they could be short wavelength transversal electromagnetic waves, similar to ultraviolet light; or (b) longitudinal electromagnetic waves (Röntgen's hypothesis); or (c) non-periodic pulses of electromagnetic radiation (Stokes' hypothesis); or (d) modified (neutral) cathode rays (Poincaré, 1897).

[3] This paper was written in 1995-1996, but it was not published. A short version, in Portuguese, was published a few years later (Martins, 2004).

2. X-RAYS AND FLUORESCENCE: POINCARÉ'S CONJECTURE

In the French Academy of Sciences, X-rays were discussed for the first time on the 20th January 1896 – just a few weeks after the publication of Röntgen's work. On this day, the first radiographs produced in Paris were also shown to the Academy. Henri Poincaré had received a pre-print directly from Röntgen, and presented a verbal account of the discovery. He remarked the importance of the new phenomenon and he became deeply interested in it. He conjectured that there could be some correlation between the emission of X-rays and the fluorescence that appeared at the glass wall of the Crookes tubes. Poincaré did not publish his hypothesis in the proceedings of the French Academy, but it appeared in a popular scientific journal (*Revue Générale des Sciences*, 30th January 1896 issue) and became widely known. In this article, Poincaré remarked:

> Therefore it is the glass that emits Röntgen rays, and it becomes fluorescent as it emits them. We may ask ourselves whether all bodies that have a sufficiently intense fluorescence wouldn't emit Röntgen's X rays, besides luminous rays, *whatever be the cause of their fluorescence*. In that case the phenomenon would not be associated to an electric cause. That is not very likely, but it is possible, and it is doubtless easy to verify. (Poincaré, 1896, p. 56)

This hypothesis will be hitherto called in this paper "Poincaré's conjecture". It was soon tested, and led to important findings (to be described below). It was the source of Becquerel's uranium research. Of course, according to our present knowledge, there is no direct relation between X-rays and luminescence, but that mistaken clue was instrumental in the discovery of several new phenomena.

Jean Becquerel, in his account of the discovery of radioactivity, ascribed to his father Henri Becquerel this conjecture:

> The day when Röntgen's first radiographs were presented to the [French] Academy of Sciences by Henri Poincaré (20th January 1896), Henri Becquerel asked his colleague where exactly was the region of emission in the tube that produced those rays. It was answered to him that the radiation came from the part of the glass wall that was stricken by the cathode rays. Henri Becquerel observed to Poincaré that this region of the glass was rendered fluorescent by the cathode rays, and both scholars immediately agreed that one should check whether other bodies besides glass, when rendered fluorescent or phosphorescent by exposure to light and not to cathode rays, would not emit a radiation similar to X-rays. Henri Becquerel soon began the research. (Jean Becquerel, 1924, p. 17)

What was the source of Jean Becquerel's version? His book provided no references. It seems, however, that he was following his father's personal narrative. In 1903 – the year he was accorded the Nobel Prize – Henri Becquerel published his only full length work of radioactivity (Becquerel, 1903a, 1903b). In this account, seven years after the beginning of his first researches on the radiation of uranium, he described the origin of his endeavour in the following words:

> In the meeting of the Academy of Sciences of the 20th January 1896, when Mr. H. Poincaré had just shown the first radiographs sent by Mr. Röntgen, I asked my colleague if it had been ascertained what was the place of emission of those rays, in the vacuum tube that produced X-rays. I was answered that the origin of the radiation was the luminous spot of the wall [of the tube] that received the cathodic flux. I cogitated at once to search whether the new emission was a manifestation of the vibratory motion that gave birth to the phosphorescence and whether all phosphorescent bodies emit similar rays. I communicated this idea and this project to Mr. Poincaré, and on the next day I began, following those ideas, a series of experiments [...]. (Becquerel, 1903a, p. 3)

Therefore, in this late publication, Becquerel ascribed to himself the origin of "Poincaré's conjecture". In Becquerel's Nobel Prize lecture (Becquerel, 1903c; translated in Samuelsson & Sohlman, 1967, pp. 52-70), he did not cite Poincaré's name:

> At the beginning of 1896, on the very day when the experiments of Röntgen and the extraordinary properties of the rays emitted by the phosphorescent wall of Crookes tubes were known at Paris, I thought of carrying out research to check whether all phosphorescent materials emitted similar rays. (Becquerel, 1903c, p. 1)

If we check Becquerel's former publications, however, we find a different account. In 1900, during the International Congress of Physics in Paris, he stated:

> The discovery of the spontaneous radiation from uranium was a consequence of the ideas born from the discovery of X-rays; [...]
>
> Discarding some hurried publications of Mr. Le Bon and Mr. Ch. Henry, whose conclusions were not verified, and that had as starting point an idea published by Mr. H. Poincaré, the first clear experiment that we find in this order of facts was due to Mr. Niewenglowski who, on the 17th February 1896, showed that some phosphorescent preparations of calcium sulphide exposed to the Sun emitted radiation that traversed black paper. [...]
>
> As to myself, since the day when I became acquainted with professor Röntgen's discovery, it came also to my mind [*il m'était également venu à l'idée*] to investigate whether the property of emitting very penetrating rays was intimately attached to phosphorescence. (Becquerel, 1900)

From this account, therefore, we could infer that Becquerel's own ideas were *independent* of Poincaré's published conjecture. We cannot be sure, however, that this weaker version is the correct one. All French physicists of the time ascribed the conjecture to Poincaré. For instance: at the meeting on 2nd March 1896 of the Academy of Sciences, Arsène d'Arsonval stated: "Fluorescent bodies emit radiation enjoying the properties of X-rays, according to the hypothesis of our colleague Mr. Poincaré" (d'Arsonval, 1896a, p. 501).

In the next meeting (9th March 1896), Louis Joseph Troost, who also detected penetrating radiation emitted by phosphorescent blende (zinc sulphide), stated: "Those results, that confirm the hypothesis of our colleague Mr. H. Poincaré and the experiments recently made by several scholars, and specially by our colleague Mr. H. Becquerel, by Mr. Niewenglowski and by Mr. Charles Henry [...]" (Troost, 1896a, p. 564).

Both d'Arsonval's and Troost's communications were read during meetings of the Academy of Sciences at which Becquerel also presented papers. It is likely that Becquerel accepted, at that time, that Poincaré was the author of this hypothesis, because he didn't claim it as his own.

It is possible to assume that Becquerel progressively changed his account and claimed the authorship of Poincaré's conjecture when he perceived the importance of radioactivity. This is consistent with Henri Becquerel's pattern of behaviour: he usually tried to ascribe to himself the contributions of other researchers (see Martins, 2000).

3. CONFIRMATIONS OF POINCARÉ'S CONJECTURE

In the weeks following the announcement of Röntgen's discovery, several papers related to X-rays were presented to the French Academy of Sciences. There was a search for different ways of producing X-rays. At the meeting on the 3rd February 1896, M. Nordon reported that a voltaic arc does not produce X-rays, but Gustave Moreau (1896) reported that they were emitted by a high voltage discharge from an induction coil, without the use of a vacuum tube (and, therefore, without the intervention of cathode rays). In the same meeting, Louis

Benoist and Dragomir Hurmuzescu reported that X-rays were able to discharge an electroscope (Benoist & Hurmuzescu, 1896) – a phenomenon that would have increasing importance in later researches.

During the following weekly meeting (10th February 1896), Charles Henry reported the first test of Poincaré's conjecture (Henry, 1896a). His paper was presented to the Academy of Sciences by Henri Poincaré himself. Charles Henry first checked whether phosphorescent zinc sulphide was able to increase the effect of X-rays produced by a Crookes tube. He concluded that it was. He covered part of a metallic object with a layer of zinc sulphide, and reported that the radiograph of this object was stronger and sharper bellow the coated region. Then, Henry proceeded to check the emission of X-rays by this phosphorescent substance when excited by light. He reported that it was possible to obtain radiographs without X-ray tubes, by covering the object with a layer of zinc sulphide and by exciting its phosphorescence burning a strip of magnesium in his laboratory. Poincaré's conjecture was confirmed.

In the next weekly meeting (17th February 1896), Gaston Henri Niewenglowski presented a confirmation of Charles Henry's results. He used another phosphorescent substance – calcium sulphide. This is the most relevant part of his report:

> I have packed an ordinary sensitive paper [photographic paper] with several layers of black or red needle paper. I placed two coins over it and covered one of the halves [of the photographic paper] with a glass plate as well as the phosphorescent powder [calcium sulphide]. After four or five hours of exposition to the Sun, the half of the sensitive paper that directly received the solar radiation remained intact and presented no mark of the coin placed upon it, thus showing that the black or red paper had not been traversed by light. The half that received the solar rays only through the phosphorescent plate was completely blackened, except for the part corresponding to one of the coins; its white silhouette was produced on black.

When I placed only one layer of thin red paper that allowed the passage of the solar rays, I observed that the portion of the sensitive paper that only received the solar radiations after their passage through the phosphorescent layer was blackened much faster than the other. (Niewenglowski, 1896, p. 385)

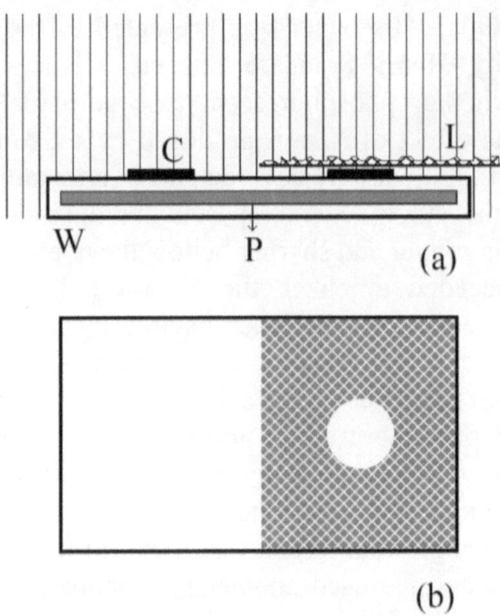

Fig. 1. A reconstruction of Gaston Niewenglowski's experiment. (a) Niewenglowski wrapped a photographic plate (P) with opaque paper (W). He placed two coins (C) over the paper, and a glass plaque with phosphorescent calcium sulphide covered half of the plaque. (b) after exposing the device to sunlight, the part of the photographic plate that was under the phosphorescent powder became dark, with a clearer and well-defined mark corresponding to the coin.

Therefore, according to the observations reported by Henry and Niewenglowski, phosphorescent materials seemed to emit X-rays, when excited by sunlight. But Niewenglowski also checked whether calcium sulphide would continue to emit X-rays when it was put in a dark place, after receiving sunlight. He concluded that this substance continued to emit penetrating radiations in the dark:

> I could also observe that the light emitted in the dark by the phosphorescent powder, previously illuminated by the Sun, was able to pass through several layers of red paper and to darken a sensitive paper that was shielded by those paper layers. (Niewenglowski, 1896, p. 386)

It was natural to check the emission of X-rays in the dark, because a sample of phosphorescent calcium sulphide keeps visibly luminous for a long time after exposition to the Sun. If X-ray emission was associated with luminescence, it was natural to expect that it could be observed also in the dark, for some time, after excitation by light.

After one more week, during the meeting of 24th February 1896, Nikolai Dmitrievich Piltchikoff reported that it was possible to increase the intensity of X-rays by placing a phosphorescent material inside the vacuum tube, at the place where the cathode rays strike the glass wall (Piltchikoff, 1896). With the older kind of vacuum tubes, it was necessary to expose photographic plates for several minutes in order to obtain a radiograph. With Piltchikof's device, the time was reduced to 30 seconds. Of course, this was another confirmation (and technological application) of Poincaré's conjecture.

All those results will strike any contemporary physicist as odd or even impossible. Nowadays we believe that luminescent bodies do not, in general, emit X-rays. Indeed, even in Röntgen's first paper it was clearly stated that X-rays could be generated when cathode rays strike *aluminium*, hence without producing any luminescence[4]. Those experiments should not

---

[4] See Röntgen's first paper (Röntgen, 1896, §§ 12-13). This was soon confirmed by Jean Perrin, who therefore denied any relation between luminescence and emission of X-rays (Perrin, 1896).

have yielded the above described results. What happened? We are unable to understand it[5].

At other countries, similar experiments were also performed. J. J. Thomson arrrived independently to Poincaré's conjecture (Thomson, 1896)[6] and tested it, with negative results:

> A very noticeable feature in the bulb producing these Röntgen rays is the phosphorescence of the glass of the bulb. I thought it therefore of interest to try if these rays were generated when the phosphorescence of the glass was produced by other means than the discharge from a negative electrode. To do this, I produced a ring discharge in an electrodeless bulb; this when the pressure of the gas is very low is accompanied by intense phosphorescence of the glass. I exposed a photographic plate protected by thick cardboard for an hour to such a bulb, but without the slightest effect. I next tried filling the bulb with oxygen, a gas which is itself made phosphorescent so as to have both the glass and the gas phosphorescent, but again a photographic plate was not affected after an hour's exposure.
>
> I also tried without success to photograph in this way by the phosphorescence excited in a screen powdered over with luminous paint, by the sparks passing between the terminals of a Ruhmkorff coil placed close to the screen.

---

[5] The increase of the photographic effect of X-rays by luminescent bodies can be partly explained. At that time, photographic plates were not very sensitive to "hard" (short wave-length) X-rays. Some substances can transform hard X-rays into "soft" (long wave-lenth) ones. Although soft X-rays have a smaller energy (and smaller penetrating power), they produce a stronger photographic effect - exactly because their absorption by matter is stronger. However, this effect cannot account for all the facts described by Charles Henry and Gaston Henri Niewenglowski.

[6] This paper was read on the 27th January 1896 and it is therefore unlikely that Thomson could have received any information about Poincaré's conjecture before he wrote it.

> It would thus appear that we can have vivid phosphorescence without any production of these rays. (Thomson, 1896)

Carey Lea, after reading a description of Charles Henry's experiment and of Poincaré's conjecture, also tested it with new substances:

> It seemed worth while to ascertain if this principle is of general application. A dilute solution of uranin was exposed to sunlight, using a large surface of solution so as to get the best effect. A short distance over the surface was placed a sensitive film protected by aluminium foil 1/10 of a millimetre in thickness and with a lead star interposed. Two hours exposure gave no result. The experiment was repeated with acid solution of quinine, with which five hours exposure gave no result. (Lea, 1896)

It is likely that Lea did not notice any effect of the uranium solution because it was dilute and spread upon a large surface. The exposure time was also too short. Notice, however, that it was *natural*, at this time, to check uranium fluorescent compounds for emission of X-rays.

In Paris, at the same meeting where Piltchikoff's work was presented, Henri Becquerel reported his first research on the emission of X-rays by phosphorescent bodies, as will be described in the next section of this paper.

4. HENRI BECQUEREL'S FIRST PAPER ON "RADIOACTIVITY"

Antoine-Henri Becquerel (1852-1908)[7] belonged to a famous scientific family. His grandfather, Antoine-César Becquerel (1788-1878) is known for his studies on electrochemistry, piezoelectricity, thermoelectricity and voltaic electricity

---

[7] He always signed his scientific works (and even his letters) as Henri Becquerel, although his full given name was Antoine-Henri.

(Knight, 1981; Henri Becquerel, 1892). Among many other works, he published a monumental review of electricity and magnetism (*Traité expérimental de l'éléctricité et du magnétisme*, in seven volumes) that remained an obligatory reference work for decades.

One of Antoine-César's sons was Alexandre-Edmond Becquerel (1829-1891), who was also a distinguished physicist (Violle, 1892; Crookes, 1892; Harvey, 1957; Gough, 1981). He initially worked with his father and applied himself to the study of several electromagnetic phenomena (electrochemistry, diamagnetism). His most important work, however, was on luminescence. He was the leading authority on phosphorescence and fluorescence of his time.

Antoine-Henri Becquerel began his scientific career in the footsteps of his father. His initial investigations were on optical phenomena – specially phosphorescence (Romer, 1981). He became familiar with ultraviolet and infrared radiations, and studied the effect of infrared radiation on the release of light by some phosphorescent substances (Becquerel, 1884a; Becquerel, 1884b; Becquerel, 1891). He studied most luminescent substances that had been collected by his father – including some uranium compounds (Becquerel, 1885). At that time, uranium compounds were an interesting subject for luminescence research, for several reasons: there were many different phosphorescent substances containing uranium; and their fluorescence was exceptionally strong. Another deeper reason for checking whether uranium compounds emitted X-rays was discussed elsewhere (Martins, 1997).

Many uranium compounds are phosphorescent or fluorescent. Among them, Edmond Becquerel had studied the nitrate, chloride, double fluoride of uranium and potassium, silicate (uranium glass), phosphate, double sulphate of uranium and potassium, etc. (Edmond Becquerel, 1859b; *idem*, 1872). Most of those uranium compounds have a very short lived phosphorescence (a few milliseconds).

With this background, it was natural that Henri Becquerel would become interested on Poincaré's conjecture and would try to check it.

Henri Becquerel's first works on "radioactivity"[8] are well known and have been translated and published several times (Romer, 1964, chapter I; Boarse & Motz, 1966, chapter 27). Those researches were published as a series of small notes in the *Comptes Rendus* of the Paris Academy of Sciences. As stated above, their starting point was his attempt to test Poincaré's conjecture.

Henri Becquerel's first research on the relation between X-rays and luminescence was presented to the French Academy on the 24th February 1896. In this report, he first acknowledged the previous studies of Charles Henry and Gaston Niewenglowski, without any criticism or reserve:

> In a previous meeting [of the French Academy of Sciences], Charles Henry reported that the intensity of the radiations that penetrate aluminium was increased when phosphorescent zinc sulphide was placed in the path of the rays that came out of a Crookes tube.
>
> Besides that, Niewenglowski discovered that commercial phosphorescent calcium sulphide emits radiations that penetrate opaque substances.
>
> This behaviour also belongs to several other phosphorescent substances, and particularly to uranium salts, which have a very short lived phosphorescence. (Becquerel, 1896a, p. 420)

Hence, in his first paper, Henri Becquerel accepted that luminescent bodies emit X-rays (or something similar to X-rays) and reported another instance of the phenomenon

---

[8] In 1896, neither the name "radioactivity" nor the corresponding concept existed, but most authors describe Becquerel's publications of that time as his "radioactivity" papers.

previously described by Charles Henry and Gaston Niewenglowski[9].

Becquerel's first experiment is remarkably similar to those of his predecessors:

> I produced the following experiment with double sulphate of uranium and potassium, of which I own some crystals that form a thin, transparent crust.
>
> A Lumière photographic plate is wrapped in two very thick black paper sheets, in such a way that the plate is not darkened even when exposed to the Sun for a whole day. A flake of the phosphorescent substance is put over the paper, outside it, and the whole is exposed to the Sun for several hours. When the photographic plate is developed, the silhouette of the phosphorescent substance appears in black in the negative. If a coin or a metallic plate with a hole is placed between the phosphorescent substance and the paper, their images will be visible in the negative.
>
> The same experiments can be repeated placing a thin glass plate between the phosphorescent substance and the paper. This excludes the possibility of any chemical action by vapours that could escape from the substance when it is heated by the rays of the Sun. It is possible to conclude from those experiments that this phosphorescent substance emits radiations that traverse a paper opaque to light and reduce silver salts. (Becquerel, 1896a)

The only relevant new aspect of Becquerel's first paper was the use of a different substance – double sulphate of uranium and potassium. The main result was similar to those of Charles Henry and Gaston Henri Niewenglowski.

5. BECQUEREL'S "DISCOVERY OF RADIOACTIVITY"

---

[9] In later works, Becquerel criticized Charles Henry's work and suggested that the observed effects were due to the pressure produced by the coin on the photographic plate (Becquerel, 1903a, pp. 4-5).

During the next meeting of the Academy of Sciences (2nd March 1896), d'Arsonval reported that he had been able to produce radiographs using a fluorescent lamp and placing a fluorescent glass over the objects that were to be radiographed (d'Arsonval, 1896a). Incidentally, the fluorescent glass he used contained an uranium salt. D'Arsonval's conclusion was that all bodies that emit greenish-yellow fluorescent light also emit radiations that are able to produce an impression on photographic plates wrapped in paper opaque to light.

At this same meeting, Henri Becquerel presented a second paper on the subject – the one that is usually described as containing the discovery of radioactivity (Becquerel, 1896b). In this article, Becquerel described new observations of the effects produced by his crystals of double sulphate of uranium and potassium. He compared the radiations of the phosphorescent substance to those produced by X-ray tubes and noticed that they had different penetration powers. He reported that the emission of penetrating radiation occurred when the phosphorescent substance received sunlight directly, or reflected by a mirror, or refracted. The part of the paper that is supposed to report the discovery of radioactivity is the following:

> I will particularly insist upon the following fact, that seem to me very important and foreign to the realm of the phenomena that one would expect to observe. The same crystalline flakes, placed together with photographic plates, in the same conditions, shielded by the same screens, but without receiving excitation by incidence of radiation and kept in the dark, still produce the same photographic impressions. This was the way I was led to make those observations: among the preceding experiments, some were prepared on Wednesday, 26th, and on Friday, 27th February; and since, in those days, the Sun appeared only intermittently, I kept the experiments that I had prepared and put the plates with their wrappings in the darkness of a furniture drawer, leaving the uranium salt flakes in their place. As the Sun did

not appear again in the following days, on the 1st March I developed the photographic plates, expecting to find very weak images. On the contrary, the silhouettes appeared with a strong intensity. I soon thought that the action should have continued in the darkness [...] (Becquerel, 1896b)

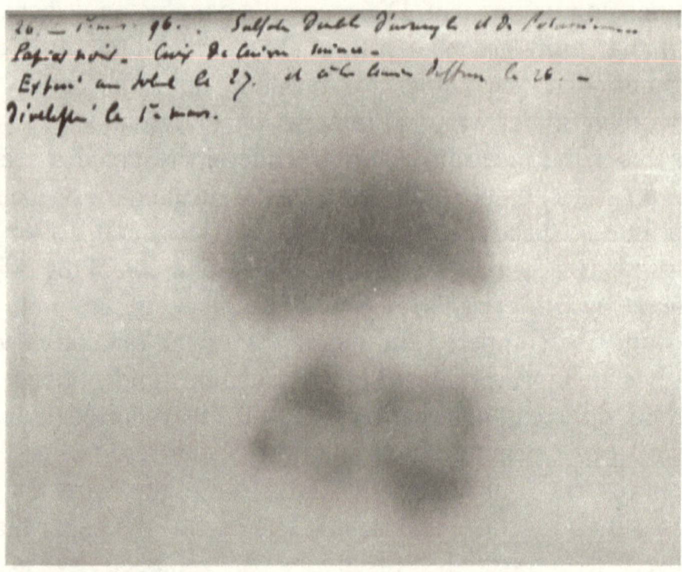

Fig. 2. One of photographs obtained by Henri Becquerel using uranium salt, when the device was kept in a drawer. The lower image was produced through a this copper cross (Becquerel, 1903a, pl. I, fig. 1)

Becquerel's original papers, published in the *Comptes Rendus* of the French Academy of Sciences, are not accompanied by any figure or plate. In much later publications, however, after radioactivity became a really important research subject, Becquerel published his early photographic plates (Becquerel, 1902a; Becquerel, 1902b; Becquerel, 1903a; Becquerel, 1903b; Becquerel, 1903c). One of them (Fig. 2) seems to correspond to Becquerel's second paper.[10]

------

[10] In Becquerel's book, this photograph (Becquerel, 1903b, plate I, fig. 1) contains a handwritten statement to the effect that it was prepared on the 26th February and developed on the 1st March 1896.

Why was this new observation unexpected? It was not the existence of penetrating radiations that was odd, but the emission of those radiations in the dark: "It seems that this phenomenon should not be ascribed to luminous radiations emitted by phosphorescence, since after one hundredth of a second those radiations become so week that they are barely perceptible". Did this show Becquerel that he was facing a completely new phenomenon, foreign to everything that physics had hitherto discovered? Not at all.

Becquerel's starting point was the (implicit) assumption that luminescent bodies would emit X-rays when they are luminous. In the case of Niewenglowski's experiment, the emission of X-rays by calcium sulphide in darkness was to be expected, because the phosphorescence of this substance is long lived. Hence, Becquerel expected a very week effect of his crystals.

The strong observed effect was unexpected, but it could be explained *according to existing knowledge about phosphorescence.* Indeed, it is only necessary to read the works of Henri Becquerel's father to find similar phenomena. The duration of the emissions of different colours of light given out by a given phosphorescent substance can be widely different. In some cases, the longer wavelengths have a longer duration (such is the case for sulphate of quinine and diamond); in other cases, the shorter wavelengths have the longer duration (chalk, Iceland spar) (Edmond Becquerel, 1859a, p. 117). For this reason, the observable colour of a phosphorescent substance, kept in the dark, usually changes with time. It could happen, therefore, that the short lived *visible* phosphorescence of the uranium crystal was accompanied by a long lived *invisible* phosphorescence with emission of penetrating radiation. This was exactly Henri Becquerel's explanation:

This seems to enhance its documental value and to establish its date. However, the statement was added much later, since in a previous publication there was nothing written on the photograph (Becquerel, 1902a, fig. 1).

> A hypothesis that naturally comes to mind is the assumption that those radiations, which effects have a strong analogy to those produced by the radiations studied by Mr. Lenard and Mr. Röntgen, could be invisible radiations emitted by phosphorescence with a persistence infinitely larger than the persistence of luminous radiations emitted by those bodies. However, the reported experiments, although not contrary to this hypothesis, do not permit us to formulate it. (Becquerel, 1896b, p. 503)

The possibility of an unknown invisible phosphorescence had already been anticipated by Henri Becquerel's father. Edmond Becquerel remarked that even in the case of substances that exhibited no observable phosphoresce in his experiments, there could be some hidden and unknown effect:

> On the other side, even supposing that the bodies are not visible in the apparatus [the phosphoroscope], one cannot state that they have received no modification, because light could excite vibrations with a speed [frequency] different from that of light rays, with a wave length greater than that of the active rays; those vibrations could give rise to heat effects and other still unknown molecular actions. (Becquerel, 1859b, p. 117)

Up to this point, therefore, Becquerel was far from suspecting he had observed anything similar to our concept of radioactivity. In a review paper on X-rays published in March 1896, Camille Raveau described the researches of Henry, Niewenglowski, Piltchikoff, d'Arsonval and Becquerel as special cases of the phenomenon predicted by Poincaré and discovered by Charles Henry (Raveau, 1896).

One week after Becquerel's famous communication of the radiation of uranium salt kept in the dark, the Academy of Sciences heard a new confirmation of Poincaré's conjecture. Louis Joseph Troost confirmed Charles Henry's experiments with phosphorescent zinc sulphide (Troost, 1896a). He obtained strong radiographic effects when its phosphorescence was excited by magnesium light. Troost referred to Niewenglowski's and Becquerel's works as investigations concerning the same phenomenon, predicted by Poincaré.

6. SILVANUS THOMPSON'S SIMULTANEOUS DISCOVERY OF HYPER-PHOSPHORESCENCE

When the scientific conditions for the rise of a discovery are ripe, it usually happens that the discovery is simultaneously made by several independent researchers. This was also the case with the "invisible phosphorescence" of uranium compounds. At the same time as Henri Becquerel, Silvanus Thompson detected this phenomenon – and interpreted it exactly in the same way as Becquerel.

Silvanus Phillips Thompson (1851-1916) is not well known today. His main work was on electricity, but he was also regarded as an authority on luminescence, at the end of the 19th century. In 1899, Lord Kelvin was interested in some aspects of the subject, and wrote a letter to Thompson, asking for help:

> Dear Thompson,
> I have looked in vain in Encyclopaedias and text-books for something that every one doesn't know regarding the phosphorescence of luminous paint, Canton's phosphorus, &c.; so as you know more than encyclopaedias and text-books put together, I apply to you. [...][11]

In Thompson's biography we find the following account of his discovery:

> During the month of February [1896] Thompson and his assistant, Mr. Miles Walker, were busily engaged in various

---

[11] Letter from Lord Kelvin to Silvanus Thompson, 10th October 1899, Cambridge University Library, Manuscripts Department, mss. Add. 7342.T164. Other manuscripts of the same Library will be hereafter be indicated as CUL.

experiments, using fluorescent substances in contact with the photographic film to hasten chemical action when stimulated by the X-rays. The materials used were finely powdered fluor-spar, sulphide of zinc, fluoride of uranium, and sundry platino-cyanides. While at work Thompson came upon an unexpected effect. He found, on developing a photographic plate, that where uranium nitrate or uranium ammonium fluoride had been used, a distinct action had taken place *through* a sheet of aluminium which is impervious to X-rays [*sic*][12]. He immediately wrote to Sir George Stokes, then President of the Royal Society, on February 26th telling him of this discovery [...] (Thompson & Thompson, 1920, p. 185)[13]

The biography does not reproduce the text of Thompson's letter. It was, however, published in Stokes' correspondence (Larmor, 1907, vol. 2, p. 495):[14]

Feb 28 1896

Dear Sir George Stokes:

I made yesterday an observation of such curious interest that I am minded to bring it before your notice. I find that if a phosphorescent substance such as sulphide of barium is exposed to ordinary white light so as to be well insolated, and brought to the shining condition it emits afterwards (and apparently also during illumination) not only ordinary light that can be cut off with an aluminium sheet, but also something else that is not cut off by aluminium, and is, in this respect at any rate, the same as the X-rays of Röntgen, that it can traverse aluminium and act on a photographic plate. If it is true that there are fluorescent (or phosphorescent) substances that deviate from your law of degradation of

---

[12] Of course, X-rays pass easily through thin plates of aluminium.

[13] There are several mistakes in this account, as will become clear afterwards. The date of the letter to Stokes is wrong, as also the content of the letter.

[14] The original letter can be found among Stokes' papers: Letter from Thompson to Stokes, 28th February 1896, CUL Add 7656.T329.

frequency (or wave-length), this would seem to present an extreme case of such deviation. But if these be Röntgen rays, then I have succeeded in manufacturing them out of common light by a sort of reversal of the process of fluorescence. Do you know of any other instance in which fluorescence or phosphorescence has been found to be reversible in its operation?

I am,

Yours most sincerely,
Silv. P. Thompson.

Notice that Thompson wrote to Stokes not because the later was the President of the Royal Society, but because he was an expert on luminescence. In this letter, Thompson said nothing about uranium compounds. It seems that the most remarkable effect was observed with sulphide of barium, since this is the only substance named in the letter. Of course, barium sulphide is not radioactive. Joseph Larmor, the editor of Stokes' correspondence, was puzzled by this letter and much time later he asked Thompson about this point.

> [...] in answer to an inquiry he now states that he had been trying various substances, including uranium nitrate, and that he had found, conclusively on Feb. 26-7, the latter was the only one to which the aluminium foil was not opaque, though black paper was transparent to others. (Larmor, 1907, vol. 2, p. 496, second footnote)

It appears from a short note published in Nature on the 12th March 1896 that Thompson was also trying different phosphorescent substances as anti-cathodes (Thompson, 1896a, p. 437)[15]. He reported that calcium sulphide, incorporated in a fusible enamel glass, "appears to form an excellent anti-kathodic [*sic*] surface for generating X-rays".

---

[15] Thompson's letter to the Editor of *Nature* was dated March 9, 1896.

According to accounts published a few months later by Thompson (Anonymous, 1896b; Thompson, 1896b; Thompson, 1896c), he tested several materials, including sulphide of calcium, fluorspar, zinc sulphide, fluoride of uranium and ammonium, and several platino-cyanides. He noticed that some of them fogged sensitive films, even when kept in the dark for a long time (fluorspar and the platino-cyanides did not exhibit this power). In the earlier experiments, there was no screen between the phosphorescent bodies and the photographic plate. Afterwards, Thompson put a thin aluminium plate between them and, in this case, only uranium nitrate and uranium ammonium fluoride affected the photographic plate. Thompson arrived at the same results that had been reached by Becquerel: some uranium compounds emitted penetrating radiations that persisted for a long time in darkness. There is no contemporary document, however, to establish the exact date of each kind of experiment.

After receiving Thompson's letter, Stokes immediately replied: "Your discovery is extremely interesting; you will I presume publish it without delay, especially as so many are now working at the X-rays" (Larmor, 1907, vol. 2, p. 495).[16]

In a copy of Stokes' reply to his letter, Thompson added a note:

> This was in reply to a letter of mine to him, telling him of my discovery that rays (which I took to be a species of hyper-phosphorescence) given off by crystals of nitrate of uranium, would pass through black paper and produce photographic effects, an observation which I thought to be discordant with his law that the rays emitted in fluorescence were always of

---

[16] A copy of the original letter can be found among Stokes' papers: Letter from Stokes to Thompson, 29th February 1896, CUL Add 7656.T330. The typewritten letter from Stokes to Thompson is kept by the Imperial College Archives, Silvanus P. Thompson's papers, letter 296 (cf. Pingree, 1967).

longer wave-length than those by which they were stimulated.[17]

A few days later, Stokes wrote again to Thompson, to give him the sad news:

> I fear that you have already been anticipated. See Becquerel, *Comptes Rendus* for Feb. 24, p. 420, and some papers in two or three meetings preceding that. (Larmor, 1907, vol. 2, p. 496)[18]

Stokes accepted that Thompson and Becquerel had discovered the same phenomenon, as may also be inferred from another letter he sent to Thompson a few months later:

> P. S. I may as well mention, in case you should not have seen it, that in the last number of the *Comptes Rendus* is a paper by Becquerel in which he mentioned that the metallic uranium shows the remarkable phenomenon which you and he discovered, independently, about 4 times as strongly as the salts of uranium he had previously used. (Larmor, 1907, vol. 2, p. 498)[19]

---

[17] CUL Add 7656.T330. This was a copy of Stokes' letter to Thompson, sent from Thompson to Larmor for publication in Stokes' correspondence. Stokes died in 1903, and shortly after that, Larmor began to collect his letters for publication. We may infer that Thompson's note was written between 1903 and 1906.

[18] A copy of the original letter can be found among Stokes' papers: Letter from Stokes to Thompson, 2nd March 1896, CUL Add 7656.T331. The typewritten letter from Stokes to Thompson is kept by the Imperial College Archives, Silvanus P. Thompson's papers, letter 297.

[19] A copy of the original letter can be found among Stokes' papers: Letter from Stokes to Thompson, 28th May 1896, CUL Add 7656.T332. The original letter from Stokes to Thompson is kept by the Imperial College Archives, Silvanus P. Thompson's papers, letter 298.

At this time, it was, of course, impossible to distinguish Becquerel's work from the researches of Charles Henry, Gaston Niewenglowski, Arsène d'Arsonval and others. Stokes probably referred specifically to Becquerel because he had for five decades been associated with the Becquerel family. On the other hand, Lord Kelvin compared Silvanus Thompson's work to that of d'Arsonval:

> I had noticed that Sylvanus Thompson's very interesting discovery had been anticipated by d'Arsonval in the Comptes Rendus of March 2.[20]

Notice that, in his first letter to Stokes, Thompson interpreted the phenomenon as a special kind of invisible phosphorescence or fluorescence that violated Stokes' law. This was not a strong objection, however, because Stokes informed Thompson that the law was not completely general: "These effects are inconsistent with a law enunciated by Stokes – but which he has since modified" (Anonymous, 1896b). The status of this law and its influence on Becquerel's thought was discussed elsewhere (Martins, 1997).

Silvanus Thompson called the phenomenon "hyper-phosphorescence" and described it in the following way:

> This phenomenon, discovered by the author independently at the same time with M. Henri Becquerel, consists in the persistent emission by certain substances, notably by metallic uranium and its salts, of invisible rays which closely resemble Röntgen rays in their photographic action, and in their power of penetrating aluminium, and of producing diselectrification. (Thompson, 1896b, p. 713)

---

[20] Letter from Lord Kelvin to Stokes, 12th March 1896. CUL Add 7656.K313. This letter was published in Wilson, 1990, vol. 2, p. 650 (letter 555).

What was the nature of those radiations? "That their [the uranium rays'] properties are intermediate between those of ultra-violet and of the Röntgen rays furnishes a strong presumption that the latter also differ only in degree, and are an extreme species of ultra-violet light" (Thompson, 1896c, p. 106).

In harmony with this interpretation, Thompson drew the same conclusion as Becquerel:

> The phenomenon of persistent emission of these invisible rays by the uranium compounds long after any electrical or luminous stimulus has ceased to be applied would seem, therefore, to bear the same relation to the transient emission of them in the Crookes tube as the persistent emission of visible light by phosphorescent bodies does to the transient emission of light by fluorescent bodies. Hence the writer ventures to give to the new phenomenon thus independently observed by M. Becquerel and by himself the name of *hyper-phosphoresce*. A hyper-phosphorescent body is one which, after due stimulus, exhibits a persistent emission of invisible rays not included in the hitherto recognized spectrum. (Thompson, 1896c, p. 106)

Henri Becquerel never mentioned Silvanus Thompson's work. Contemporary physicists ascribed small importance to Thompson's contribution, perhaps because Becquerel's results were published a few months before Thompson's reports. Historians of science seldom refer to his contribution.[21] However, even among French writers, the name "hyperphosphorescence" coined by Thompson was widely accepted and used to describe the phenomenon studied by Henri Becquerel (Guillaume, 1897, pp. 131-135). It was Marie Curie

---

[21] One exception is Lawrence Badash, who cited his name (Badash, 1965, p. 63, footnote 29), but provided no information about Thompson's work.

(not Becquerel) who first criticized the name and the concept underlying it:

> I will call *radioactives* the substances that emit Becquerel rays. The name *hyperphosphorescence* that had been proposed for the phenomenon seems to me to convey a wrong idea about its nature. (Curie, 1899)

7. DO PHOSPHORESCENT ANIMALS EMIT X-RAYS?

Henri Becquerel's researches should be analysed in their historical context. After the discovery of X-rays, scientific periodicals were flooded both by papers concerning the radiation emitted by Crookes' tubes and by communications of new kinds of radiations.

Among the many researches triggered by the discovery of X-rays, one of the most curious was the search for X-ray emitted by animals. The motivation for this search was, once again, Poincaré's conjecture: if the production of X-rays is related to luminescence, then luminescent animals should also emit X-rays.

On 18th April 1896, Raphael Dubois presented to the Biology Society the first evidence of emission of penetrating radiation by a phosphorescent animal (Dubois, 1896a). He studied a bivalve mollusc (*Pholad*) that emitted "a beautiful bluish phosphorescence, that offers to the eye some analogy to that of mineral bodies". Dubois observed that the luminous organ of the mollusc was able to have an effect on a photographic plate, through black paper, after an exposition of 15 hours (without the black paper, it took 12 hours to obtain a photograph). It was also possible to obtain photographic effects through cardboard and thin wood (a few millimetres thick). In this case, it took 18 hours to obtain the effect. He also tried aluminium, but the results were not very clear. His conclusion was:

Those are, certainly, encouraging endeavors that I will repeat with luminous microbes, but they are not sufficient to demonstrate in a rigorous way the existence of X rays among living bodies, because ordinary radiations could perhaps pass through thin bodies in an amount that cannot be noticed by the eye, but sufficient to impress in a long time a sensible plate, by accumulative action. (Dubois, 1896a, p. 385)

One month later (9th May 1896), Dubois presented the results of his study of luminous microbes. The abstract of his work described positive results:

I have exposed, above liquid cultures of luminous photo-bacteria, photographic plates wrapped in two or even in three sheets of the paper used to protect those plates from ordinary light. One coin was placed between the paper sheets. The whole was placed in the most complete darkness. After twenty-four hours of exposition, and after development, the contours of the coin and of the open vessel that contained the culture were clearly distinguished: the space comprised between those two contours was neatly impressed. (Dubois, 1896b, p. 479)

Although this text was careful enough to avoid concluding that those living beings emitted X-rays, the title of his work ("X-rays and luminous microbes") clearly shows Dubois' belief.

Dubois' work had no impact. A few months later, however, there appeared another similar work that was seriously discussed by the scientific community.

On the 5th August 1896, Han'ishi Muraoka[22] sent to the *Annalen der Physik und Chemie* a paper on the light emitted by a kind of firefly (Muraoka, 1896). Muraoka had studied Becquerel's work on the penetrating radiation emitted by fluorescent compounds of uranium, and he conjectured that light emitting insects could also produce penetrating rays similar to X-rays.

---

[22] Han'ichi Muraoka (1853-1929) was a physics professor at Kyoto.

Muraoka used photographic plates upon which he placed a card with a cross-shaped hole. Upon the card he placed thin plates of different metals (aluminium, copper, zinc, etc.) and the whole was wrapped in three or four folds of black paper. The prepared plate was then exposed during two or three days to the radiation emitted by about 300 fireflies[23]. The card between the metal and the photographic plate was intended to avoid contact effects that had already been studied by MacIntyre and that were observed to produce effects on the photographic plates[24]. Muraoka reported a positive effect, and he studied the penetration of the rays through different metals and substances. The rays did not affect fluorescent screens, and produced no electric discharge. Muraoka described evidence that the rays could be reflected, but he has found no definite evidence of refraction or polarisation.

Those rays had some strange properties. The effect was stronger directly under a cardboard than at the places where the cardboard had a hole. The cardboard seemed to produce some kind of "suction effect" upon the rays, analogous to the effect of iron for magnetic lines of force. It also seemed that only upon filtration by cardboard, paper or metal, the radiation emitted by the fire-flies was able to produce effect upon photographic plates.

Muraoka concluded that the rays emitted by his insects was similar to Becquerel's rays, that he called "Becquerel's fluorescent rays" [*Becquerel'schen Fluorescenzstrahlen*].

Muraoka's work called the attention of several researchers[25]. Among them, Silvanus Thompson was specially interested in

---

[23] The insects used by Muraoka appeared in Kyoto around the middle of June and were named by him (in German) as "Johanniskäfer", that means *St. John's beetle*. The name was not adequate, and the fireflies used by Muraoka were later identified as *Luciola vitlicollis* and *Luciola picticollis* (see Lungo, 1897, p. 130).

[24] MacIntyre's work is described in Anonymous (1896a, p. 379).

[25] See, for instance, the abstract published in *The Electrician* **38**: 238, 1896, where Muraoka's paper is described as "the sensation of the

the phenomenon, and he reported that Dawson Turner had also observed that glow-worms emit rays that can pass through aluminium (Thompson, 1897, p. 126).[26]

Independently of Muraoka, Charles Henry had described a related phenomenon (Henry, 1896b). He put a few glow-worms over photographic plates wrapped in black paper. When the plates were developed, he observed dark tracks reproducing the tracks of the worms upon the plates. However, Charles Henry did not give much attention to this effect.

Muraoka sent a copy of his paper on "Das Johanniskärferlicht" to Stokes, who replied to him[27] on the 18th March, 1897. Stokes found Muraoka's work "extremely interesting" and suggested to him some new experiments.

Stokes had once observed fireflies and had a hypothesis concerning the production of light by those insects: "I could not help supposing that the light arose in some way from an electric discharge, made at the will of the animal, as in the case of electric fishes, though how the discharge, if there were one, produced the light, I could not tell". Therefore, Stokes was led to think that at least part of the effect observed by Muraoka could be due to an electric discharge. But he also accepted the possibility of "beetle rays", that is, radiation emitted by the insects and that could traverse black paper.

Stokes supposed that "beetle rays" could not traverse metal plates, but that they could produce a slight electric charge upon the surface of the metal (such as happens with X-rays or Becquerel rays). This electric charge could flow through the photographic plate and produce the observed effects. For this reason, the effect was stronger directly under a metal plate, as Muraoka had described. Accordingly, Stokes suggested that

---

current number of Wiedemann's *Annalen*". Abstracts were also published in the *American Journal of Science* IV, **3**: 151-152, 1897, and in several other scientific journals.

[26] It seems that Turner never published those observations.

[27] Letter from Stokes to Muraoka, 18 March 1897, CUL Add. 7656.M759.

Muraoka connected the metal plates to the Earth, or placed a thin mica foil between the metal plates and the photographic plate.

Muraoka replied to Stokes[28] in a letter dated 12th October 1897. He tried to analyse the light emitted by the fireflies, but could obtain no result since he used a glass prism (that is opaque to ultraviolet rays). He tried to detect electric currents in the beetles, and connected to the Earth the metal plates, with no positive effect.

He reported that the use of mica produced an inversion of the darker and lighter parts of the image on the photographic plate. He was led to think that besides the direct effect of the penetrating beetle rays there should be a second cause. In this letter, Muraoka stated: "Several experiments I made led me to assume vapours of bodies used in the experiments as the second cause". In some experiments with wood, he believed that the resin also produced photographic effects: "When the action of the beetle rays surpasses that of the resin vapour then is the action at the softer part stronger and in the other case the denser part appears darker".

Muraoka still believed that the fireflies emitted penetrating rays, but he could not find new effects, and for this reason he sent for publication in the *Annalen der Physik und Chemie* only a discussion of the effect of vapours upon photographic plates.

Muraoka's second paper (Muraoka & Kasuya, 1898) was received for publication in the *Annalen der Physik und Chemie* on the 24th November 1897. In this paper, Muraoka and Kasuya recall that several researchers had also described penetrating rays from luminous insects[29]. Stokes' letter to Muraoka was acknowledged and part of it was translated.

---

[28] Letter from Muraoka to Stokes, 12 October 1897, CUL Add. 7656.M760.

[29] They cite the works of Charles Henry, R. Dubois, K. Shimada and D. Turner (*apud* Silvanus Thompson).

The main influence studied in the second paper was the effect of water. In the first paper, Muraoka had already described that it was necessary to keep the insects slightly wet so that they could live for several days. In the second paper, Muraoka reported that the darkening of the photographic paper was different according as the insects were more or less dry. This observation led him to investigate the effect of damp and vapours upon photographic plates. He noticed that the presence of several substances (such as coffee) would darken the photographic plates. Some metals, such as zinc, cadmium and magnesium also affected photographic plates.

Although there is no final conclusion in this paper concerning the previous researches on the radiation of fireflies, most readers must have concluded that the former results were spurious and that the observed effects were all due to vapour emitted by several substances. In a review of experiments related to "Becquerel rays", Stewart (1898) commented Muraoka's researches:

> Much interest was excited but a short time ago over a supposed invisible radiation from glow worms. The discoverer [Muraoka] has lately announced that the effect was in some way due to moisture, it being necessary to keep the glow worms wet. Moistened paper gave the same effect that the glow worms had.

Notice, however, that this was not, Muraoka's belief. In his second letter to Stokes, he still maintained that the action of vapours was a "second cause", that could be weaker or stronger than the direct effect of the beetle rays.

8. DOES SUGAR EMIT PENETRATING RAYS?

Of course, the early search for bodies emitting X-rays or other radiations was not limited to phosphorescent animals. One of the most famous claims of 1896 was Gustave le Bon's discovery of "black light" – a radiation emitted by common flames, that could traverse thick opaque bodies and thin metal plates[30]. In this case, as in several other episodes of the time, there was an intricate mixture of interesting new phenomena and mistakes.

In a letter to Gustave le Bon, written on the 6th June 1896, Auguste Lumière complained about the multitude of "new radiations" discovered after the X-rays:

> Everyday someone submits to us some cases of this kind; we have by the hundreds cases of a partial or total clouding of photographic plates, of impressions produced in extraordinary circumstances, without intervention of light, or without apparent intervention. In most of these cases which are submitted to us constantly, we have almost always been able to find the cause of these phenomena, after a study – albeit sometimes very lengthy – of each of them.[31]

The Lumière brothers, who produced most of the photographic plates used in those researches, carefully checked and dismissed many early claims – such as the supposed emission of penetrating radiations by electric arcs and flames – as due to heat or penetration of common light (Lumière & Lumière, 1896a). On the 24th February, the same day when Henri Becquerel presented his first "radioactivity" paper, the Lumière brothers warned again that many different agents could affect photographic plates: mechanical pressure, contact with several substances, heat, penetration of light through the boxes or wrappings that contained the plates, etc. (Lumière & Lumière, 1896b). Similar warnings were published by René Colson, who added to the Lumières' list the action of water vapour, and infrared and ultraviolet radiations that can sometimes pass through considerably opaque bodies (Colson, 1896).

---

[30] This interesting episode will not be discussed here; see Nye (1974).
[31] Letter from Auguste Lumière to Gustave le Bon, 6 June 1896, cited by Nye (1974).

Several researchers have looked for materials that emitted penetrating radiations. RenéColson and Henri Pellat described the action of some metals (zinc, magnesium, cadmium, steel) on photographic plates (Colson, 1896; Pellat, 1896). Colson ascribed the effect to metal vapour, but Pellat suggested that the observed effects could be due to penetrating radiations similar to those of uranium.

F. McKissick (1897), inspired by Becquerel's work, reported that the following substances produced photographic effects through the cover of a plate holder: lithium chloride, barium sulphide, calcium sulphate, quinine chloride, quinine sulphate, calcium nitrate, sugar, chalk, glucose, sodium tungstate, sterein, uranium acetate, and ammonium phospho-molybdate. Both glass and metal were opaque to the radiation emitted by those substances, while cardboard and wood were "very transparent". The most active of all substances tried by McKissick was sugar: "I have succeeded in taking a fairly clear negative through 2½ in. of wood with sugar".

There were some unbelievable aspects in McKissick report: "Generally more than one image of an object was produced, although the object was in direct contact with the sensitive plate. In one negative there are four images of one half-dollar, in another two images of a key, the images being at right angles to each other". Of course, this is only possible if the object moved over the photographic plate during the exposition.

In 1897, Russell Russell began his researches repeating some of Becquerel's experiments but soon obtained anomalous results: when a perforated zinc screen was put between the active substance and the photographic plate, the part of the photographic place below the hole was less effected than the part under the metal plate (Russell, 1897). Repeating the experiment without uranium compounds, Russell noticed that zinc alone would darken the photographic plate. The action of zinc was observable even when paper, parchment or rubber screens were interposed between the metal and the photographic plate.

Other metals were observed to produce similar effects: mercury, magnesium, cadmium, zinc, nickel, aluminium, pewter, fusible metal (an alloy of lead, bismuth and tin), lead, bismuth, tin, cobalt, antimony. Other metals, such as iron, gold and platinum, produced no observable effect. He reported that wood was active, as also charcoal, straw and silk. The ink used in some newspapers was also very active.

In some cases, Russell conjectured that vapours emitted by the active substance could be the cause of the observed effect. However, it was difficult to accept that this could be the case with metals. Further experiments (Russell, 1898), however, confirmed that the active metals had the property of giving off (even at ordinary temperature) some kind of vapour which could affect photographic plates. This vapour was able to pass through thin sheets of paper, gelatine, celluloid, etc., and could be carried by a current of air. A last series of experiments (Russell, 1898) led to the conclusion that the effect was not due to metallic vapour, but to hydrogen peroxide produced at the surface of metals by reaction with air and moisture.

All those instances show how difficult it was, in 1896-97, to understand what meaning should be ascribed to spots on photographic plates. In some of his experiments, Becquerel had only wrapped his photographic plates in black paper – and the observed effect could be due to anything, from X-rays to heat and hydrogen peroxide. Only from 1898 those pitfalls were avoided by the use of electrical methods.

Gerhard Schmidt was led to discover that thorium and its compounds emit radiations similar to those of uranium when he studied the several substances that seemed able to darken photographic plates (Schmidt, 1898). Schmidt was aware of the works of Muraoka, Henry, Russell and others. He observed that the uranium radiations were different from the "radiations" of all other substances, because only uranium radiations rendered the air an electric conductor. He investigated several other substances, and was lucky enough to find out that thorium compounds also darken photographic plates *and* produce electric conduction in the air. Marie Curie was also led to the simultaneous discovery of the radiation of thorium by the same method (Curie, 1898).

The discovery of a second metal that produced radiations similar to those of uranium led, as is well known, a new impulse to the research of what we call radioactivity. But the fundamental finding was not the discovery of the radioactivity of thorium: it was the new electrical research method. It was only after Schmidt and Curie began to use electrical conduction as the criterion for recognition of radiations, in 1898, that it was possible to conclusively dismiss the supposed emission of penetrating radiations by metals, glow worms and sugar.

9. FINAL COMMENTS

Viewed "in the lump" among other researches concerning penetrating radiation emitted by several bodies, in the period 1896-1897, Becquerel's work was just one among several strange reported phenomena. Most of this early research was guided by Poincaré's conjecture. During all this period, Becquerel himself made no effort to draw a distinction between his own research and that of other people who had verified Poincaré's conjecture. Except for the case of Gustave le Bon's "black light" – that was violently attacked by Becquerel – he never criticised the works that described other penetrating radiations.

The discovery of X-rays excited the imagination of scientists and laymen alike, and it was followed by a large amount of speculative activity[32]. Among the several suggestions about the nature of X-rays, Poincaré's conjecture was particularly fertile. It was easy expose photographic plates to luminescent bodies and to look for something similar to X-rays. On the other hand, hypotheses such as Röntgen's suggestion of longitudinal ether

---

[32] According to Oliver Lodge, general doubts about accepted knowledge and speculative activity are usually produced by new discoveries (Lodge, 1912).

waves were much more difficult to test and led to a limited number of publications (Thomson, 1896; Kelvin, 1896).

The nature and the process of production of X-rays was unknown, and even the empirical recognition of X-rays did not follow clear rules. That explains the widespread confusion between X-rays and many other effects that nowadays we would classify as spurious or as representing different phenomena. In this period the scientific community was, as a whole, highly uncritical and scientific periodicals accepted for publication reports that nowadays we would describe as ridiculous.

Among other issues, this episode raises the problem of evaluation of experimental research: was it possible, in 1896-1897, to distinguish Becquerel's work from other reported findings, and to assess their relative scientific values? In principle, yes. Any old-fashioned manual of scientific method would tell us that one should test the reliability of the experimental techniques themselves[33]. In practice, however, there was no systematic precaution concerning replicability, control of confounding factors, variation of conditions and observation techniques, etc. The lack of rigour must have been perceived by skilled scientists – but they probably did not pay much attention to the many claims of new radiations and effects following Röntgen's discovery.

Before 1898 – when the emission of radiations from thorium was found – not much attention was paid to Becquerel's research. He also must have thought that it was not a very interesting subject – just a new kind of invisible phosphorescence – and turned his attention to another subject: the Zeeman effect.

---

[33] See, for instance, Mario Bunge (1967). In the recent literature, the possibility of establishing criteria for acceptance of empirical knowledge has been usually denied. Some authors, however, maintain the existence of objective experimental criteria. See Franklin (1986), Woodward & Bogen (1988), Culp (1995).

It was not Becquerel who called the attention of the world to a new, strange phenomenon. It was due to Schmidt's and Curie's works that the radiations emitted by uranium (and thorium) were clearly distinguished from other reported effects.

It was Marie Curie and not Becquerel who rejected the name "hyperphosphorescence" and coined the word "radioactivity", dismissing Poincaré's hypothesis and calling the attention of the scientific world to a new class of phenomena. Marie Curie was responsible not only for the word "radioactivity", but also for the establishment of radioactivity as a new research field. It was mainly due to Marie Curie's work, after the discovery of the radioactivity of thorium, polonium, and radium, that the subject became widely known and discussed, and the research of radioactivity became fruitful and was detached from the swarm of strange effects that arose around X rays.

ACKNOWLEDGEMENTS

This work was produced while the author was a visiting scholar of the Department of History and Philosophy of Science, University of Cambridge, and a visiting fellow of Wolfson College. The author is grateful to the São Paulo State Research Foundation, Brazil (FAPESP) for the support of this research.

BIBLIOGRAPHIC REFERENCES

[ANONYMOUS]. The Röntgen rays. *Nature* **53**: 377-380, 1896 (a).

[ANONYMOUS]. Physics at the British Association. *Nature* **54**: 565-566, 1896 (c).

BADASH, Lawrence. "Chance favors the prepared mind": Henri Becquerel and the discovery of radioactivity. *Archives Internationales d'Histoire des Sciences* **18** (70): 55-66, 1965 (a).

BECQUEREL, Edmond. Recherches sur divers effets lumineux qui résultent de l'action de la lumière sur les corps. *Annales de Chimie et de Physique* [3] **55**: 5-119, 1859 (a).

BECQUEREL, Edmond. Recherches sur divers effets lumineux qui résultent de l'action de la lumière sur les corps. Composition de la lumière émise (troisième mémoire). *Annales de Chimie et de Physique* [3] **57**: 40-124, 1859 (b).

BECQUEREL, Edmond. Mémoire sur l'analyse de la lumière émise par les composés d'uranium phosphorescents. *Annales de Chimie et de Physique* [4] **27**: 539-579, 1872.

BECQUEREL, Henri. Spectres d'émission infra-rouges des vapeurs métalliques. *Comptes Rendus Hebdomadaires des Séances de l'Académie des Sciences de Paris* **99**: 374-376, 1884 (a).

BECQUEREL, Henri. Détermination des longueurs d'onde des raies et bandes principales du spectre solaire infra-rouge. *Comptes Rendus Hebdomadaires des Séances de l'Académie des Sciences de Paris* **99**: 417-420, 450, 1884 (b).

BECQUEREL, Henri. Relations entre l'absorption de la lumière et l'émission de la phosphorescence dans les composés d'uranium. *Comptes Rendus Hebdomadaires des Séances de l'Académie des Sciences de Paris* **101**: 1252-1256, 1885.

BECQUEREL, Henri. Sur les différentes manifestations de la phosphorescence des minéraux sous l'influence de la lumière ou de la chaleur. *Comptes Rendus Hebdomadaires des Séances de l'Académie des Sciences de Paris* **112**: 557-563, 1891 (a).

BECQUEREL, Henri. La chaire de physique du Muséum. *Revue Scientifique* **49**: 673-8, 1892.

BECQUEREL, Henri. Sur les radiations émises par phosphorescence. *Comptes Rendus Hebdomadaires des Séances de l'Académie des Sciences de Paris* **122**: 420-421, 1896 (a).

BECQUEREL, Henri. Sur les radiations invisibles émises par les corps phosphorescents. *Comptes Rendus Hebdomadaires des Séances de l'Académie des Sciences de Paris* **122**: 501-503, 1896 (b).

BECQUEREL, Henri. Sur le rayonnement de l'uranium et sur diverses propriétés physiques du rayonnement des corps radio-actifs. Vol. 3, pp. 47-78, in: GUILLAUME, Charles-Édouard ; POINCARÉ, Lucien (eds.). *Rapports Présentés au Congrès International de Physique réuni a Paris en 1900*. 3 vols. Paris: Gauthier-Villars, 1900.

BECQUEREL, Henri. Sur la radio-activité de la matière. *Notice of the Proceedings of the Meetings of the Members of the Royal Institution of Great Britain* **17**: 85-94, 1902 (a).

BECQUEREL, Henri. On the radioactivity of matter. *Annual Report of the Board of Regents of the Smithsonian Institution* 197-206, 1902 (b).

BECQUEREL, Henri. Recherches sur une propriété nouvelle de la matière – activité radiante spontanée ou radioactivité de la matière. *Mémoires de l'Académie des Sciences de l'Institut de France* **46**: 1-360, 1903 (a).

BECQUEREL, Henri. *Recherches sur une propriété nouvelle de la matière: activité radiante spontanée ou radioactivité de la matière*. Paris: Firmin-Didot, 1903 (b).

BECQUEREL, Henri. Sur une propriété nouvelle de la matière, la radio-activité. *Les Prix Nobel* **3**: 1-15, 1903 (c).

BECQUEREL, Jean, La radioactivité et les transformations des éléments. Paris: Payot, 1924.

BENOIST, Louis & HURMUZESCU, Dragomir. Nouvelles propriétés des rayons X. *Comptes Rendus Hebdomadaires des Séances de l'Académie des Sciences de Paris* **122** (5): 235-236, 1896.

BOARSE, Henry A. & MOTZ, Lloyd (eds.). *The world of the atom*. 2 vols. New York: Basic Books, 1966.

BUNGE, Mario. *Scientific research*. 2 vols. Berlin: Springer-Verlag, 1967.

COLSON, René. Rôle des différentes formes de l'énergie dans la photographie au travers des corps opaques. *Comptes Rendus Hebdomadaires des Séances de l'Académie des Sciences de Paris* **122**: 598-600, 1896 (a).

COLSON, René. Action du zinc sur la plaque photographique. *Comptes Rendus Hebdomadaires des Séances de l'Académie des Sciences de Paris* **123**: 49-51, 1896 (b).

CROOKES, William [W. C.]. Edmond Becquerel. *Proceedings of the Royal Society of London* **51**: xxi-xxiv, 1892.

CULP, Sylvia. Objectivity in experimental inquiry: breaking data-technique circles. *Philosophy of Science*, **62** : 438-458, 1995.

CURIE, Marie Sklodowska. Rayons émis par les composés de l'uranium et du thorium. *Comptes Rendus Hebdomadaires des Séances de l'Académie des Sciences de Paris* **126**: 1101-3, 1898.

CURIE, Marie Sklodowska. Les rayons de Becquerel et le polonium. *Révue Générale des Sciences* **10**: 41-50, 1899.

D'ARSONVAL, Arsène. Observation au sujet de la photographie à travers les corps opaques. *Comptes Rendus Hebdomadaires des Séances de l'Académie des Sciences de Paris* **122**: 500-501, 1896.

DUBOIS, Raphael. Les rayons X et les êtres vivants. *Comptes Rendus Hebdomadaires des Séances et Mémoires de la Société de Biologie* [10] **3**: 384-385, 1896 (a).

DUBOIS, Raphael. Les rayons X et les microbes lumineux. *Comptes Rendus Hebdomadaires des Séances et Mémoires de la Société de Biologie* [10] **3**: 479, 1896 (b).

FRANKLIN, Allan. *The neglect of experiment*. Cambridge: Cambridge University Press, 1986.

GLASSER, Otto. Wilhelm Conrad Röntgen and the early history of the Röntgen rays. London: John Bale, Sons & Danielsson, 1933.

GOUGH, Jerry B. Becquerel, Alexandre-Edmond. Vol. 1, pp. 555-556, in: GILLISPIE, Charles C. (ed.). *Dictionary of scientific biography*. New York: Charles Scribner's Sons, 1981.

GUILLAUME, Charles-Édouard. *Les rayons X et la photographie a travers les corps opaques*. 2. ed. Paris: Gauthier-Villars et Fils, 1897.

HARVEY, E. Newton. *A history of luminescence from the earliest times until 1900*. Philadelphia: The American Philosophical Society, 1957.

HENRY, Charles. Augmentation du rendement photographique des rayons Roentgen par le sulfure de zinc phosphorescent. *Comptes Rendus Hebdomadaires des Séances de l'Académie des Sciences de Paris* **122**: 312-314, 1896 (a).

HENRY, Charles. Utilité, en radiographie, d'écrans au sulfure de zinc phosphorescent; émission, par les vers luisants, de rayons traversant le papier aiguille. *Comptes Rendus Hebdomadaires des Séances de l'Académie des Sciences de Paris* **123**: 400-401, 1896 (b).

JAUNCEY, George Eric Macdonnell. The birth and early infancy of X-rays. *American Journal of Physics* **13**: 362-379, 1945.

KELVIN, Lord. On the generation of longitudinal waves in ether. *Proceedings of the Royal Society of London* **59**: 270-272, 1896.

KNIGHT, David M. Becquerel, Antoine-César. Vol. 1, pp. 557-558, in: GILLISPIE, Charles C. (ed.). *Dictionary of scientific biography*. New York: Charles Scribner's Sons, 1981.

LARMOR, Joseph (ed.). *Memoir and scientific correspondence of the late Sir George Gabriel Stokes*. 2 vols. Cambridge: Cambridge University Press, 1907.

LEA, Mathew Carey. Röntgen rays not present in sunlight. *American Journal of Science* [4] **1**: 363-364, 1896. Reproduced in: *Philosophical Magazine* [5] **41**: 528-530, 1896.

LODGE, Oliver. The discovery of radioactivity, and its influence on the course of physical science (Becquerel memorial lecture). *Journal of the Chemical Society* **101**: 2005-2032, 1912.

LUMIÈRE, Auguste & LUMIÈRE, Louis. Recherches photographiques sur les rayons de Röntgen. *Comptes*

Rendus Hebdomadaires des Séances de l'Académie des Sciences de Paris **122** (7): 382-383, 1896 (a).

LUMIÈRE, Auguste & LUMIÈRE, Louis. A propos de la photographie à travers les corps opaques. *Comptes Rendus Hebdomadaires des Séances de l'Académie des Sciences de Paris* **122**: 463-464, 1896 (b).

LUNGO, Carlo del. Fire-fly light. *Nature* **57**: 130, 1897.

MARTINS, Roberto de Andrade. Becquerel and the choice of uranium compounds. *Archive for History of Exact Sciences* **51** (1): 67-81, 1997.

MARTINS, Roberto de Andrade. Los errores experimentales de Henri Becquerel. Pp. 267-274, in: GARCÍA, Pío; MENNA, Sergio H.; RODRÍGUEZ, Víctor (eds.). *Epistemología e Historia de la Ciencia. Selección de Trabajos de las X Jornadas*. Facultad de Filosofia y Humanidades. Vol. 6. Córdoba: Universidad Nacional de Córdoba, 2000.

MARTINS, Roberto de Andrade. Hipóteses e interpretação experimental: a conjetura de Poincaré e a descoberta da hiperfosforescência por Becquerel e Thompson. *Ciência & Educação* **10** (3): 501-516, 2004.

MCKISSICK, A. F. Becquerel rays. *The Electrician* **38**: 313, 1897.

MOREAU, Gustave. De la photographie des objets métalliques à travers des corps opaques, au moyen d'une aigrette d'une bobine d'induction, sans tube de Crookes. *Comptes Rendus Hebdomadaires des Séances de l'Académie des Sciences de Paris* **122**: 238-9, 1896.

MORGAN, Augustus. *A budget of paradoxes*. London: Longmans, Green and Co., 1872.

MURAOKA, Hanishi. Das Johanniskäferlicht. *Annalen der Physik und Chemie* [2] **59**: 773-81, 1896.

MURAOKA, Hanishi & KASUYA, M. Das Johanniskäferlicht und die Wirkung der Dämpfe von festen und flüssigen Körpern auf photographische Platten. *Annalen der Physik und Chemie* [2] **64**: 186-92, 1898.

NIEWENGLOWSKI, Gaston Henri. Sur la proprieté qu'ont les radiations émises par les corps phosphorescents, de traverser certains corps opaques à la lumière solaire, et sur les expériences de M. G. Le Bon, sur la lumière noire. *Comptes Rendus Hebdomadaires des Séances de l'Académie des Sciences de Paris* **122**: 385-386, 1896.

NYE, Mary Jo. Gustave Le Bon's black light: a study in physics and philosophy in France at the turn of the century. *Historial Studies in the Physical Sciences* **4**: 163-195, 1974.

PELLAT, Henri. Sur la vaporisation des métaux à la temperature ordinaire. *Comptes Rendus Hebdomadaires des Séances de l'Académie des Sciences de Paris* **123**: 104-105, 1896.

PERRIN, Jean. Origine des rayons de Röntgen. *Comptes Rendus Hebdomadaires des Séances de l'Académie des Sciences de Paris* **122** (12): 716-717, 1896.

PILTCHIKOFF, Nikolai Dmitrievich. Sur l'émission des rayons de Röntgen, par un tube contenant une matière fluorescente. *Comptes Rendus Hebdomadaires des Séances de l'Académie des Sciences de Paris* **122**: 461, 1896.

PINGREE, Jeanne. *Silvanus Phillips Thompson, F. R. S. List of correspondence and papers in the Imperial College Archives.* London: Imperial College, 1967.

POINCARÉ, Henri. Les rayons cathodiques et les rayons Roentgen. *Revue Générale des Sciences* **7**: 52-9, 1896.

POINCARÉ, Henri. Les rayons cathodiques et les rayons Roentgen. *Revue Scientifique* [4] **7**: 72-81, 1897.

RAVEAU, Camille. Les faits nouvellement acquis sur les rayons de Roentgen. *Revue Générale des Sciences* **7**: 249-53, 1896.

ROMER, Alfred. *The discovery of radioactivity and transmutation.* New York: Dover, 1964.

ROMER, Alfred. Becquerel, [Antoine-] Henri. Vol. 1, pp. 558-561, in: GILLISPIE, Charles C. (ed.). *Dictionary of scientific biography.* New York: Charles Scribner's Sons, 1981.

RÖNTGEN, Wilhelm Conrad. Über eine neue Art von Strahlen (Vorläufige Mittheilung). *Sitzunsberichte der physikalisch-medicinischen Gesellschaft zu Würzburg* (9): 132-141, 1895.[34]

RÖNTGEN, Wilhelm Conrad. On a new kind of rays. Trad. Arthur Stanton. *Nature* **53** (1369): 274-276, 1896

RUSSELL, William James. On the action exerted by certain metals and other substances on a photographic plate. *Proceedings of the Royal Society of London* **61**: 424-433, 1897.

RUSSELL, William James. Further experiments on the action exerted by certain metals and other bodies on a photographic plate. *Proceedings of the Royal Society of London* **63**: 102-112, 1898.

RUSSELL, William James. On hydrogen peroxide as the active agent in producing pictures on a photographic plate in the dark. *Proceedings of the Royal Society of London* **64**: 409-419, 1899.

SAMUELSSON, Bergt; SOHLMAN, Michael (eds.). Nobel lectures including presentation speeches and laureates' biographies. Physics. 1901-1921. Amsterdam: Elsevier, 1967.

SARTON, George. The discovery of X-rays. *Isis* **26**: 349-369, 1937.

SCHMIDT, Gerhard C. Ueber die von den Thorverbindungen und einigen anderen Substanzen ausgehende Strahlung. *Annalen der Physik und Chemie* [2] **65**: 141-151, 1898. Reprinted in: *Verhandlungen der physikalische Gesellschaft nach Berlin*, **17**: 14-16, 1898.

---

[34] Reproduced in *Annalen der Physik und Chemie* [3] **64** (1): 1-11, 1898. English translations: On a new kind of rays. Trad. Arthur Stanton. *Nature* **53** (1369): 274-276, 1896; On a new form of radiation. *The Electrician* **36** (13): 415-417, 1896. French translation: Une nouvelle espèce de rayons. *Révue Générale des Sciences Pures et Appliquées* **7**: 59-63, 1896.

STEWART, Oscar M. A résumé of the experiments dealing with the properties of Becquerel rays. *Physical Review* **6**: 239-251, 1898.

THOMPSON, Jane Smeal; THOMPSON, Helen G. *Silvanus Phillips Thompson, his life and letters*. London: T. Fisher Unwin, 1920.

THOMPSON, Silvanus P. The Röntgen rays. *Nature* **53**: 437, 1896 (a).

THOMPSON, Silvanus P. On hyperphosphorescence. *Report of the 66th Meeting of the British Association for the Advancement of Science* [**66**]: 713, 1896 (b).

THOMPSON, Silvanus P. On hyperphosphorescence. *The London, Edinburgh and Dublin Philosophical Magazine and Journal of Science* [série 5] **42**: 103-107, 1896 (c).

THOMPSON, Silvanus P. Fire-fly light. *Nature* **56**: 126, 1897.

THOMSON, Joseph John. Longitudinal electric waves, and Röntgen's X rays. *Proceedings of the Cambridge Philosophical Society* **9**: 49-61, 1896 (b).

TROOST, Louis Joseph. Sur l'emploi de la blende hexagonale artificielle pour remplacer les ampoules de Crookes. *Comptes Rendus Hebdomadaires des Séances de l'Académie des Sciences de Paris* **122**: 564-6, 1896 (a).

TROOST, Louis Joseph. Observation à l'occasion de la communication de M. H. Becquerel. *Comptes Rendus Hebdomadaires des Séances de l'Académie des Sciences de Paris* **122**: 694, 1896 (b).

VIOLLE, Jules. L'oeuvre scientifique de M. Edmond Becquerel. *Revue Scientifique* **49**: 353-360, 1892.

WATSON, E. C. The discovery of X-rays. *American Journal of Physics* **13**: 281-291, 1945.

WILSON, David B. (ed.). *The correspondence between Sir George Gabriel Stokes and Sir William Thomson, Baron Kelvin of Largs*. 2 vols. Cambridge: Cambridge University Press, 1990.

WOODWARD, James; BOGEN, James. Saving the phenomena. *The Philosophical Review* **97**: 303-352, 1988.

DID NIEPCE DE SAINT-VICTOR DISCOVER RADIOACTIVITY?

Roberto de Andrade Martins

Abstract: It is usually accepted that Henri Becquerel discovered radioactivity in 1896. However, up to the beginning of 1898, nobody (including Becquerel himself) interpreted the phenomenon he studied as anything similar to our current concept of radioactivity. Becquerel did observe effects due to uranium radiation, but he interpreted it as due to an invisible phosphorescence. His early works are full of mistakes concerning the properties of the phenomenon he studied. This article compares Becquerel's work to the researches of Abel Niepce de Saint-Victor. Forty years before Becquerel's investigations, Niepce had detected a persistent radiation emitted by uranium nitrate in the dark. Niepce's interpretation (an invisible phosphorescence) is similar to Becquerel's and different from our concept of radioactivity. It is likely that he made experimental mistakes and intermingled different phenomena. However, if mere contact with effects of a phenomenon, without a clear understanding of its nature and properties, can count as the discovery of that phenomenon, one could argue that it was Niepce – not Becquerel – who discovered radioactivity. The paper uses this example to discuss the concept of discovery of a scientific phenomenon.
Keywords: radioactivity; scientific discovery; history of physics; Becquerel, Henri; Saint-Victor, Niepce

MARTINS, Roberto de Andrade. *Historical Essays on Radioactivity*. Extrema: Quamcumque Editum, 2021.

1. INTRODUCTION

In 1903, Antoine-Henri Becquerel[1] (1852-1908) was accorded the Nobel Prize, together with Pierre and Marie Curie, "in recognition of the extraordinary services he has rendered by his discovery of spontaneous radioactivity" (Wasson, 1987, p. 70; Samuelsson & Sohlman, 1967, p. 45). At that time, as today, it was generally accepted that Henri Becquerel discovered radioactivity in 1896, when he observed that a compound of uranium was able to darken photographic plates wrapped in black paper, kept in a drawer. Historical accounts sometimes stress that this was a lucky chance discovery, or discuss the background of the discovery to show that Becquerel was looking for something and found it (Jauncey, 1946; Romer, 1964; Romer, 1970; Badash, 1965a; Badash, 1965b; Badash, 1966). But the discovery itself is never challenged.

It seems that Henri Becquerel can be regarded as the discoverer of radioactivity because:

a) He was regarded as the discoverer of radioactivity by himself and by the scientific community in the early 20th century;

b) Nobody else claimed to be the discoverer of radioactivity, at that time;

c) Historians of science did not find anyone else who discovered radioactivity before Becquerel.

There is, however, a dangerous implicit assumption in this argument: the premise that *someone* must have discovered radioactivity. Contrary to the usual belief, this article will claim that in the common-sense meaning of the word "discovery", Becquerel did not discover radioactivity: the discovery of this phenomenon was the result of a gradual and collective effort beginning with Röntgen's work on X-rays and culminating with Rutherford and Soddy's theory of transmutation of the elements. If, however, one wishes to interpret "discovery" in a way that would allow us to maintain that Becquerel discovered radioactivity, the same criterion could be applied to maintain

---

[1] For a biography, see Romer (1981).

instead that radioactivity was discovered in 1857 by a gentleman called Abel Niepce de Saint-Victor.

This article is not intended as an example of the historiographical vice of "priority chasing" (May, 1975). The main body of this paper will be devoted to the elucidation of Henri Becquerel's and Niepce de Saint-Victor's researches, and their scientific context, in order to elucidate the long path and the difficulties in discovering (and understanding) a new phenomenon.

2. BECQUEREL'S EARLY CONTRIBUTION TO THE STUDY OF RADIOACTIVITY

In 1896 and in the early 1897, Henri Becquerel published a series of papers that nowadays are regarded as the first description of radioactivity. The word "radioactivity" did not exist at that time, and Becquerel's interpretation of the observed phenomena was different from our current interpretation; however, for simplicity, those articles will be hereafter called Becquerel's "radioactivity" papers.

Let us summarize the conclusions of those publications:

2.1. *A penetrating radiation – 24th February 1896*

Becquerel detected with a photographic plate penetrating radiation emitted by phosphorescent samples (double sulfate of uranyl and potassium) under sunlight (Becquerel, 1896a). The experiment was a variant of those reported by Charles Henry (1896) and Gaston Niewenglowski (1896), who had detected penetrating radiations emitted by other phosphorescent substances (zinc sulphide and calcium sulphide). The motivation of those investigations was Poincaré's conjecture that luminescent substances could emit X-rays (Poincaré, 1896).[2]

---

[2] For a description of the context of Becquerel's research, see the previous paper in this volume: MARTINS, Roberto de Andrade. A pool of radiations: Becquerel and Poincaré's conjecture.

2.2. *Emission of radiation in the dark – 2nd March 1896*

Becquerel observed that the radiation emitted by his samples was able to pass through thin metal plates. He also observed that his samples of double sulphate of uranyl and potassium emitted penetrating radiation even when kept in the dark. He remarked that the visible phosphorescence of those samples is very short (1/100 s). He proposed the hypothesis that the phenomenon was produced by radiation similar to X-rays emitted by phosphorescence, with a persistence much larger than that of visible phosphorescence of the same substance (Becquerel, 1896b).

2.3. *Properties of the radiation and test of other substances – 9th March 1896*

Becquerel reported that the crystals of double sulphate of uranyl and potassium were able to discharge an electroscope. He did not interpret this phenomenon as due to gas ionization and supposed that it was due to the direct action of radiation upon the charged bodies. He noticed that the emission of penetrating radiation persisted after a few days in darkness and conjectured that the phenomenon was similar to calorescence. He also reported that the radiation was reflected by a polished metal mirror and by glass; and that it could be refracted by glass. In this paper, Becquerel described a series of tests with different phosphorescent bodies, searching for other substances that emitted similar penetrating rays. According to Becquerel, all phosphorescent uranium salts, and two calcium sulphide samples, emitted penetrating radiation (Becquerel, 1896c).[3]

2.4. *Emission of radiation by all uranium salts – 23rd March 1896*

---

[3] With this paper, Becquerel started the publication of a series of wrong results. A discussion of Becquerel's errors is presented in: MARTINS, Roberto de Andrade. Becquerel's experimental mistakes, in this volume.

In this paper, Becquerel confirmed the refraction of uranium radiation by glass, using a prism. He reported that his calcium sulphide samples did not emit penetrating radiation in a second series of experiments (an effect, he remarked, similar to what had been observed by Troost with zinc sulphide). Becquerel observed that all uranium compounds he tested emitted penetrating radiation – even those compounds that were not luminescent. The emission of radiation seemed to increase when the samples were illuminated with a strong light.[4] Becquerel concluded that the phenomenon seemed due to an invisible phosphorescence that was not directly linked to visible luminescence (Becquerel, 1896d).

2.5. *Comparison between radiation of uranium compounds and X-rays – 30th March 1896*

Both X-rays and the radiation emitted by uranium compounds produce photographic effects, discharge electroscopes and pass through matter. Becquerel noticed, however, that their penetrating powers were different. Besides, X-rays could not be polarized, but Becquerel reported that the radiation emitted by uranium compounds could be polarized by tourmalines.[5] He measured a decrease of intensity of radiation emission a few hours after exposure to light. Becquerel suggested that X-rays and the radiation emitted by uranium compounds are different and that both might be emitted by X-ray tubes (Becquerel, 1896e).

2.6. *Emission of radiation by metallic uranium – 18th May 1896*

In this paper, Becquerel first summarizes his previous results: uranium salts emit radiation for several months, with a

---

[4] According to current knowledge, the emission of radiation by radioactive bodies cannot be increased by light.

[5] That was another mistake, that confirmed Becquerel's opinion that the radiation emitted by uranium compounds was similar to ultraviolet radiation.

small decrease of intensity; emission can be excited by strong light; the radiation can be refracted and reflected; this radiation can pass through thin metal plates and discharges an electroscope; and all uranium salts emit this radiation, irrespective of their visible luminescence properties. Due to this last property, Becquerel was led to test metallic uranium. He observed that it also emitted the penetrating radiation and the emission was stronger than that of uranium compounds. He concluded that uranium was the first example of a metal that presented a phenomenon of invisible phosphorescence (Becquerel, 1896f).

2.7. *Persistence of uranium rays – 23rd November 1896*

For six months, Becquerel published no new research on this subject. In his last publication of 1896, he stressed that the radiation he had studied was different from X-rays because the former could be reflected and refracted as light, and proposed the name "uranium rays". He reported that emission of radiation was still intense after several months, although there was a small weakening. The article called attention to the difficulty in explaining the source of energy in the phenomenon. In the same paper, Becquerel confirmed for uranium rays Joseph John Thomson's finding that X-rays produce electric discharge by ionizing air (Becquerel, 1896g).

2.8. *Electric discharge produced by uranium rays – 1st March and 12th April 1897*

In the two last articles of this series, Becquerel described further studies of the electric discharge produced by uranium radiation (Becquerel, 1897a; Becquerel, 1897b). There is no new remark about the nature of the radiation or about the process of its production. Becquerel described that the samples kept in darkness for one year continued to emit penetrating radiation "with a just decreasing intensity" (*avec une intensité à peine décroissante*) (Becquerel, 1897b, p. 803).

2.9. *After 1897*

It is clear that up to 1897 Becquerel's work is a mixture of correct and wrong observations, together with usually wrong interpretations. At that time, he was far from understanding the phenomenon he was studying as we understand it.

After the publication of those "radioactivity" papers, Becquerel turned his attention to the study of Zeeman's effect.[6] Only one year after Marie Curie's and Gerhard Schmidt's discovery of the radioactivity of thorium (Curie, 1898; Schmidt, 1898) Becquerel returned to the study of radioactivity (Becquerel, 1899).[7] Meanwhile, most of Becquerel's mistakes had been corrected, the name "radioactivity" was coined (by Marie Curie), and radioactivity was recognized as a new and important phenomenon.

3. THE STATUS OF BECQUEREL'S WORK TOWARDS THE BEGINNING OF 1898

What did the scientific community think that Becquerel had found, before the discovery of the radioactivity of thorium in 1898?

Several review papers[8] on uranium (or Becquerel's) rays were written from 1896 to 1898. The first report that dealt exclusively with Becquerel's rays was written by Georges Sagnac (1896). It was written in April and it was published in May 1896. Short reviews appeared in articles and books primarily concerned with X-rays (Thomson, 1896; Thompson,

---

[6] Becquerel published five papers on this subject in the *Comptes Rendus*, from 1897 to the beginning of 1899.

[7] The first paper of the second series of radioactivity studies was presented to the French Academy of Sciences on the 27th March 1899.

[8] We are including among review papers only those articles that are primarily aimed to the description of previous work and not to publication of original research results.

1897; Guillaume, 1897; Poincaré, 1897). A detailed review was published by Oscar Stewart (1898).[9]

All those reviews told essentially the same story. They accepted without discussion Becquerel's experimental results and described his discovery of an *invisible phosphorescence*, or *hyperphosphorescence*:

a) All uranium compounds (and even metallic uranium) and some other phosphorescent substances (calcium sulphide, zinc sulphide) emit an invisible radiation that can pass through thin plates of metal and other substances; this radiation can be detected by its photographic effects.

> a.1) Some other substances (metals, wood, paper) had also been found to emit invisible penetrating radiation.[10]
>
> a.2) Uranium radiation can discharge an electroscope, as X-rays do.

b) The radiation studied by Becquerel can be reflected, refracted and polarized; it is therefore a kind of invisible light (similar to ultraviolet rays).

> b.1) This was the main distinction between those rays and X-rays, because the latter could not be reflected, refracted and polarized.

c) The penetrating radiation is emitted during a long time when the active substances are kept in darkness, but the intensity of radiation increases when the substance is excited by sunlight.

> c.1) In the case of blende and calcium sulphide, the effect gradually disappears and cannot be revivified; in the case of substances containing uranium, the radiation could be slightly increased by exposition to sunlight and other radiations.

---

[9] Published in February 1898, this article was written without awareness of the recent discovery of the radioactivity of thorium. The same author also published another review two years later (Stewart, 1900).

[10] At that time, there was not a clear distinction between all those phenomena. For details, see my previous paper in this volume.

d) Because of all the previous properties, the phenomenon is similar to common (visible) phosphorescence and can be called *invisible phosphorescence* (or *hyperphosphorescence*).

Of course, that does not correspond to our present-day knowledge of radioactivity. It was not just an *incomplete* knowledge of the phenomenon we call radioactivity – it was a *erroneous* partial account of the phenomena. Indeed, nowadays we accept that:

a') All uranium compounds (and even metallic uranium) *but no other common phosphorescent substances* (calcium sulphide, zinc sulphide) emit an invisible radiation that can pass through thin plates of metal and other substances; this radiation can be detected by its photographic and electrical effects.

> a'.1) Metals, wood, paper, etc. *do not* emit invisible penetrating radiation similar to the radiation emitted by uranium compounds.

b') The radiation studied by Becquerel *cannot* be reflected, refracted and polarized (at least not in the way Becquerel said it could); therefore, the correct conclusion at that time was that it is *not* a kind of invisible light similar to ultraviolet rays.

c') The penetrating radiation is emitted during a long time when the active substances are kept in darkness, *and the intensity of radiation does not increase* when the substance is excited by sunlight or other kinds of radiation.

d') Because of all the previous properties, the phenomenon is *not* similar to common (visible) phosphorescence and therefore it *should not* be called *invisible phosphorescence* (or *hyperphosphorescence*).

The previously described wrong account (a-d) was accepted as true by Henri Becquerel, in the early 1898, and most evidence for that mistaken view had been provided by himself. All the above described corrections (a'-d'), together with a lot of new information, were produced in the period 1898-1899, as the result of contributions of many different researchers, specially Gerhard Carl Schmidt, Julius Elster, Hans Geitel, Marie and Pierre Curie, and Ernest Rutherford, without any important contribution coming from Becquerel.

In the early 1900, Oscar Stewart published a second review paper on Becquerel rays (Stewart, 1900). He remarked that "The importance of the subject of Becquerel rays has increased almost beyond expectation during the past year or two" and then proceeded to describe the completely new view on radioactivity that had been reached at that time. In 1900, at the International Congress of Physics, Marie and Pierre Curie presented an essentially correct (although incomplete) description of radioactivity:

> We shall therefore only recall that uranium rays, or *Becquerel rays*, are characterized by the following properties: they have rectilinear propagation; they act upon photographic plates as light, but in an extremely weak degree; they can pass through screens of different materials, but only if they are very thin; they are neither reflected, nor refracted, nor polarized; when they pass through a gas, they render it a weak electrical conductor.
>
> Uranium radiation is spontaneous and constant; it is not maintained by any known exciting cause; it seems insensible to changes of temperature and illumination. (Curie & Curie, 1900, p. 79)

Becquerel, however, had not arrived to this view of radioactivity.

4. IF BECQUEREL DISCOVERED RADIOACTIVITY, WHEN DID HE DO IT?

It is generally accepted by historians of science that Becquerel discovered radioactivity in the first months of 1896. Some of them associate the discovery with the second "radioactivity" paper and the observation of emission of radiations by an uranium salt kept in darkness:

On Sunday, 1 March 1896, Henri Becquerel developed a set of barely exposed photographic plates – and discovered the phenomenon of radioactivity. (Badash, 1966, p. 267)

Jean Becquerel (Henri's son) has also pointed out Becquerel's second "radioactivity" paper as the relevant turning point:

Therefore, it was useless to expose the flakes to sunlight and to produce their phosphorescence; the emission of the new radiation was produced in the drawer, protected from all known exciting radiation; that emission was *spontaneous*, and seemed to defy the principle of conservation of energy because at that time there was no reason to think that matter could be an energy reservoir. Radioactivity had been discovered. (Jean Becquerel, 1924, p. 20)

Other historians prefer to describe the discovery of emission of radiations by metallic uranium as the culminating point of Becquerel's contribution:

With this last announcement, on 18 May, Becquerel's discovery of radioactivity was complete [...]. (Romer, 1981, p. 559)[11]

At another place, Romer also associated Becquerel's paper of 18th May 1896 to the completion of the discovery of radioactivity:

Seven weeks more went by and the discovery was complete. Uranium always gave out the penetrating rays, whether it was in fluorescent or non-fluorescent crystals, whether in the light or in the dark, whether dissolved in water or isolated in Moissan's pure and uncombined metal. (Romer, 1964, p. 18)

---

[11] Notice that this was not Becquerel's last paper.

Generally, historians do not associate the discovery of radioactivity with Becquerel's *first* experiment. Indeed, at that time he observed for the first time the photographic effect of a radioactive substance, but he was far from understanding that he was facing a new phenomenon. For that reason, a date is usually chosen that seems related to some fundamental insight on the nature of the new phenomenon.

Once it was even claimed that Becquerel discovered radioactivity *several years before* 1896. According to Bertrand (1946), towards the end of 1893 or beginning of 1894, Henri Becquerel had already observed that a piece of pitchblende was able to darken a nearby photographic plate wrapped in black paper, kept in a drawer. Becquerel was not able to understand the phenomenon, and Bertrand suggested to him a chemical explanation: maybe the photographic plate had been affected by vapours emanated from the mineral. He suggested that Becquerel should try whether pitchblende could act upon the photographic plate through a thin tin sheet or paper impregnated with lead acetate.

> Some days afterwards, Henri Becquerel came again to see me and informed that my explanation was not correct, but, he added, *I have found it.* I was not indiscreet and did not ask him what was it; the following discoveries gave me the elucidation that I wanted and, at the same time, they revealed us that the fortuitous impression of a spot produced by pitchblende upon a photographic plate was the first observation of radioactivity made by Henri Becquerel. (Bertrand, 1946, p. 699)

Becquerel himself never claimed that he observed this radiation effect with pitchblende before 1896. It could happen that Bertrand's recollection was not correct.

Even those who do not attach a precise date to the discovery accept the Henri Becquerel was the discoverer of radioactivity. After all, that was the reason why he was nominated for the Nobel Prize:

The Royal Academy of Sciences of Sweden decides, on the 12th November 1903, to confer the Nobel Prize in *physics* of this year to

HENRI-ANTOINE BECQUEREL

for his *discovery of the spontaneous radio-activity*

and also to

PIERRE CURIE

and to

MARIE-SKLODOVSKA CURIE

for their *works concerning the radiation phenomena discovered by* HENRI BECQUEREL.[12]

When – and in what sense – did Becquerel discover radioactivity? The two questions are deeply interrelated.

The phenomenon we now call radioactivity is a process of spontaneous transformation of some unstable atomic nuclei, with emission of specific kinds of radiation (α, β and γ). The discoverer of the phenomenon we call radioactivity should therefore be someone who found out that:

a) some atomic nuclei can be transformed into other different nuclei;

b) this transformation is spontaneous (that is, the source of energy is internal, and no external phenomenon triggers the transformation);

c) the transformation has a speed that falls exponentially with time, with a characteristic half-time for each kind of atomic nucleus, that cannot be changed by external influences (temperature, chemical reactions, etc.);

d) there are different kinds of radiation emitted by those nuclei (alpha particles, equal to Helium nuclei; beta particles, equal to

---

[12] *Les Prix Nobel en 1903* (Stockholm, 1906), 2. Notice that Becquerel's name was Antoine-Henri, not Henri-Antoine. Also, the hyphen between Marie and Sklodowska is wrong.

electrons; and gamma rays, an electromagnetic radiation similar to high-energy X-rays).[13]

Not a single one of those components of our current concept of radioactivity can be ascribed to Becquerel. One can safely state that none of those aspects of radioactivity had been ascertained by Becquerel or anyone else before 1900 (except, perhaps, its spontaneity). Indeed, each of those aspects of radioactivity was established through a long line of experimental research combined with theoretical work. Many people contributed bringing together some parts of the puzzle – and sometimes with parts that did not fit the puzzle. Around 1903, on the contrary, the knowledge about natural radioactivity was close to what we accept nowadays. If some single name should be credited for this discovery, a likely candidate would be Ernest Rutherford, to whom is due a considerable fraction of our knowledge of radioactivity.

It would be very odd, however, to call Rutherford the discoverer of radioactivity. After all, he was not the first one to observe effects due to radioactivity. But what ground is there to give the title to Becquerel? Was his contribution a singular, necessary step in the development of radioactivity research?

Let us play with counterfactuals. Suppose Henri Becquerel had died before 1896. Would there be any delay in the discovery of radioactivity? Probably not. Independently of Becquerel, in the early 1896, other researchers were looking for penetrating radiations emitted by phosphorescent substances, and it was natural to try uranium compounds.[14] Without the contribution of Becquerel, it could even occur that other people – perhaps

---

[13] Nowadays we know, of course, that the emission of beta particles is accompanied by neutrinos. In the case of artificial nuclides, instead of the usual beta rays (electrons) the atom may emit positrons (positive electrons).

[14] Independently of Becquerel, Silvanus Thompson and Lea Carey searched for X-rays emitted by uranium compounds. For details, see my previous paper on this volume.

Silvanus Thompson – would make *correct* observations and notice that uranium radiation is not reflected, refracted or polarized. In that case, the development of the field would have been faster than it really was.

Yes, but it *was* Becquerel and no one else who observed the first effects of radioactivity, wasn't he? Not exactly. Before the discovery of the radioactivity of thorium, Becquerel (as most other scientists) believed that he had found a new kind of phosphorescence. That is: he supposed that some substances (anything containing uranium), after receiving energy from sunlight, slowly emitted an invisible, penetrating electromagnetic radiation similar to ultraviolet light. He did neither suggest that there was any subatomic transformation involved in what he studied, nor that the transformation was spontaneous, nor did he understand what kind of radiation was emitted by uranium. What he observed was not radioactivity, but hyperphosphorescence.

Someone might say: "Well, Becquerel didn't provide the correct *interpretation* of radioactivity, but after all he discovered the phenomenon".

Did he? If we analyse the details of Becquerel's work, it is possible to perceive that not only his interpretation, but even the *facts* described by him were not correct. The empirical properties of Becquerel's phenomenon do not correspond to the known properties of radioactive bodies – they correspond to his *opinion* about the phenomenon.

This case may be compared to Columbus' discovery of America: he had a wrong opinion about the size of the Earth, and for that reason he thought that it would be easy to arrive to India traveling westwards; he arrived to a place that we nowadays call "America", but he thought he had arrived to China or India (Mentré, 1905). Of course, empirical evidence was against his belief: the language, customs, dresses and physical appearance of the inhabitants was different from what was expected. Animals and plants were also different from those that were known to inhabit South Asia. Even when the accumulated evidence was clearly against his belief, however, he retained his interpretation. He never suspected he had discovered a new continent. In which sense did Columbus discover America? Only in the sense that he was the first European of his time to reach the place we call America and to announce his travels to the Old World.

Sometimes the "discovery" of a phenomenon is reduced just to this: a first contact or a first observation of something, even if the interpretation was wrong. Is it possible, in that sense, to ascribe to Becquerel the discovery of radioactivity? In that case, should we associate the discovery with his first experiment?

Let us suppose that Becquerel's first contact with radioactivity did happen as Bertrand told it. In that case, should we change the date of the discovery of radioactivity and say that Becquerel discovered radioactivity about the end of 1893 or beginning of 1894?

Finally, what is the *minimum* accomplishment someone must have done in order to be credited with the discovery of radioactivity? If we want to maintain that Becquerel discovered radioactivity in 1896, it will be necessary to reduce the "discovery of radioactivity" to this: some substances (specially those that contain uranium) emit invisible radiations that can pass through thin opaque bodies and produce photographic effects. Although Becquerel thought at first that this effect was produced by sunlight, he also noticed that the emission of radiation would continue for a long time in darkness.

Now, if that is enough to establish the discovery of radioactivity, then radioactivity was discovered 40 years *before* Becquerel, by Niepce de Saint-Victor.

5. NIEPCE DE SAINT-VICTOR'S EXPERIMENTS ON LIGHT STORAGE

Hitherto, historians of science have not seriously considered the possibility that Henri Becquerel had been anticipated in the discovery of radioactivity. Badash remarked that uranium compounds had been in use for decades, "but there had been no hints that uranium was steadily emitting unseen, penetrating radiations".

> This is perhaps unusual in itself. Once a significant discovery is made, numerous priority claims are often lodged. To my knowledge, there have been no public claims for the prior discovery of radioactivity, and only one simultaneous assertion by Silvanus P. Thompson. (Badash, 1965, p. 63, footnote 29)

However, not long after the publication of Henri Becquerel's early "radioactivity" papers, there was a claim that Niepce de Saint-Victor had discovery the invisible, penetrating uranium radiation in the period 1857-1861. In his very popular book *L'évolution de la matière*, Gustave le Bon charged Becquerel with plagiarism:

> It was at the same time that Mr. Becquerel published his first researches. Repeating the forgotten experiments of Niepce de Saint-Victor and making use of uranium salts, as he [Niepce] had done, he [Becquerel] has shown, as the former one had already shown, that those salts give off in the dark some radiations that can affect photographic plates. Continuing for a longer time than his predecessor the experiments, he observed that the emission seemed to persist indefinitely.
>
> What are those radiations? Always under the influence of the ideas of Niepce de Saint-Victor, Becquerel initially believed that they amounted to what Niepce called "stored light", meaning a kind of invisible phosphorescence; and to prove it, he set up experiments which were described at length in the *Comptes Rendus de l'Académie des Sciences* and which led him to believe that the radiations emitted by uranium are refracted, reflected and polarized. (Le Bon, 1906, pp. 35-36)

Claude-Félix-Abel Niepce de Saint-Victor (1805-1870) was a cousin or nephew (Larousse, 1865-1876, vol. 11, p. 999) of the famous Joseph Nicéphore Niepce – one of the originators of photography. Influenced by his relative's work, he began to develop photographic experiments in 1846, that led him to the discovery of the first process for producing negatives on glass using a film of albumen to hold the sensitive compound (Gibson, 1923, p. 23; Potonniée, 1936, pp. 218-222).[15] Among his several researches on photography and light, there was a very interesting study on what could be called an invisible phosphorescence by common materials.[16]

> Does a body, after being strocken by light or put under the Sun, keep in the dark some impression of that light? That is the problem that I tried to solve by photography. Phosphorescence and fluorescence of bodies are well known; but, as far as I know, nobody ever did experiments such as those that I am going to describe.

In his first experiments,[17] Niepce de Saint-Victor used a printed paper. It was first kept in darkness for several days. Then, half of the paper was covered with an opaque screen, and the other half was exposed to sunlight during 15 minutes or more. Afterwards, in a dark room, the whole printed paper was applied to a photographic plate. After one day, the plate was developed and showed a negative copy of the part of the print that had received the light.[18]

---

[15] His first publications: Niepce de Saint-Victor (1847; 1848a; 1848b, 1850a; 1850b).

[16] His results were first published as short communications to the Paris Academy of Sciences, and afterwards as a full report (Niepce de Saint-Victor, 1857-1867; Niepce de Sain-Victor, 1861). Some of his communications were also published (in full or as extracts) in other journals.

[17] Meeting of 16th November 1857 of the Paris Academy of Sciences.

[18] The observed effect could be partially explained by Colson's finding, four decades later, that contact with dry ink affects photographic plates: it produces an oxydation effect that turns the plate insensible to light, at the points of contact (Colson, 1896). A

Niepce de Saint-Victor reported that all kinds of papers produced the same effect, although in different degrees. Similar effects were produced by wood, ivory, parchment, marble, chalk, porous porcelain, cotton, and other substances. Metals, wood charcoal, vitrified porcelain and glass produced no effect.

The effect seemed not be produced by heat, since black or dark paper produced no effect upon the photographic plate, although dark surfaces become hotter than white paper under sunlight.

At a first sight, the list of substances that Niepce de Saint-Victor described as emitting invisible radiations seems very odd. Let us, however, recall the knowledge of his time about phosphorescence.

There are some phosphorescent substances that can shine in the dark for a long time – even several hours. This effect was first observed with some jewels, and in 1604 the so-called Bologna phosphor (barium sulphide) was first described (Edmond Becquerel, 1859, p. 9). Afterwards, several other strongly phosphorescent bodies were found, such as the Canton phosphor (calcium sulphide), strontium sulphide and a kind of calcium fluoride called *chlorophane*. In the 19th century, systematic search showed that many other substances exhibited a short lived phosphorescence (lasting from a few seconds to a fraction of a second). Among those substances, it is relevant to cite chalk, sugar, paper (Edmond Becquerel, 1859, pp. 10-11), and several other organic substances, such as tartaric acid, lactose, teeth, silk, etc (*ibid.*, pp. 22-23). Therefore, most of the active substances described by Niepce de Saint-Victor were not entirely devoid of phosphorescent properties. Edmond Becquerel stated:

> The phenomenon of phosphorescence by insolation is much more general than is generally thought, [...] We shall see that a very large number of bodies give rise to effects of

---

similar suggestion had already been made much before (Malone, 1862).

the same order as the alkaline earth sulphides, and, as I have proved, certain substances that do not present emission of light after insolation, keep nevertheless the impression due to the action of the radiation, but for a time too short to allow the effect to be observed in ordinary circumstances. (Edmond Becquerel, 1859, p. 12)

It could happen, however, that those substances emitted some non-visible radiation for a long time after their visible phosphorescence had died out.

By interposing plates of different substances between the printed paper and the photographic plate, Niepce de Saint-Victor observed that the effect did not traverse glass, mica, crystal, etc. A print covered by gelatin or collodion could be reproduced by this process, but varnish or glue prevented the effect.

Niepce de Saint-Victor examined whether the effect was due to the direct contact between the print and the photographic plate. Even when they were separated by a few millimetres, the effect still occurred.

Among several curious observations, he described the following experiment. The inner surface of an iron tube was covered with white paper and exposed to the rays of the Sun for about one hour. If it was hermetically closed immediately after exposition to light, "it will keep during an indefinite time the power of radiation that was communicated to it by the sunlight" (Niepce de Saint-Victor, 1861, p. 37). Hence, it seemed possible to store light in a can.

Niepce de Saint-Victor also made some experiments with fluorescent substances.

A drawing on white paper, made with a solution of quinine sulphate, one of the most fluorescent of known bodies, exposed to the Sun and applied to a sensible [photographic] paper, is reproduced in black with a greater intensity than the white paper that constitutes the basis of the drawing. (Niepce de Saint-Victor, 1861, p. 38)

He remarked that the effect was not due to chemical action: when the quinine drawing was not exposed to light, it produced no effect upon the photographic paper. In his second paper, Niepce de Saint-Victor remarked that the effect was stronger replacing the quinine by a solution of tartaric acid or uranium nitrate. The effect was also observed when there was a distance of two or three centimeters between the drawing and the photographic paper, if the line of the drawing was sufficiently thick.

In his second communication on this subject (1st March 1858), Niepce de Saint-Victor described a new kind of experiment.

> One takes a sheet of paper that was kept several days in darkness; one covers it with a photographic negative on glass or paper; one exposes it to the rays of the Sun during a longer or shorter time, according to the intensity of light, and bring it to darkness; one takes out the negative and treat [the paper] with a solution of silver nitrate. One sees, in a short time, an image that can be fixed by washing in pure water. If one wants to obtain a faster and stronger image, the paper sheet should be impregnated beforehand with a substance [...] with a stronger power of storing the light activity. One very efficient substance of this kind is a water solution of uranium nitrate, obtained by treating uranium oxide with diluted nitric acid, or by dissolving in water uranium nitrate crystals. (Niepce de Saint-Victor, 1861, pp. 39-40)

This observation, as will be seen below, led to the development of new photographic materials, at the time.

A similar, but weaker effect, was obtained replacing uranium nitrate by tartaric acid. A few other substances also produced similar effects: citric acid, oxalic acid, aluminum sulphate, iron citrate, etc.

The experiment of light storage in metal tubes was also reproduced with paper containing uranium nitrate and tartaric acid, and he obtained stronger effects than before.

I expose to the light of the Sun a sheet of cardboard strongly impregnated with two or three layers of a solution of tartaric acid or uranium salt. After insolation I cover with this cardboard the inner part of a long and thin white iron tube. I hermetically close the tube and I notice that the cardboard impresses a sensible paper prepared with silver chloride after a very long time delay, the same as on the first day. [...] The experiment only succeeds once, that is, it seems that the light escapes completely from the cardboard, and, in order to obtain a second image, it is necessary to have recourse to a second insolation. (Niepce de Saint-Victor, 1861, p. 43)

Niepce de Saint-Victor remarked that the uranium salts are strongly fluorescent, but tartaric acid exhibits no fluorescence whatever. Therefore, the effect seemed independent of phosphorescence or fluorescence.

The phenomenon described by Niepce de Saint-Victor may seem to us unbelievable, but it was not completely different from other known phenomena. Let us recall that when a phosphorescent substance is exposed to light and brought to a dark room, it will shine during some time, but its luminosity will decrease and after a longer or shorter time the substance will seem to have lost all its phosphorescent light. However, there are several phosphorescent substances that can shine again after becoming dark, if they are heated. They can store light energy during a long time. This phenomenon is, of course, different from the one described by Niepce de Saint-Victor, but there are some similarities: in both cases there is some kind of hidden phosphorescence that can remain for a long time in a substance that is not shining any more. Besides, as will be seen below, the active substances produced stronger and faster effects when they were heated – exactly as it occur in calorescence.

In a later communication, Niepce added that the effect of the tube with uranium nitrate or tartaric acid was the same as the first day, *even after several months* (Niepce de Saint-Victor, 1867). The conclusion of the second communication was:

> The experiments described in this Memoir prove, I think, in the most evident way, that light communicates a real activity to some substances stricken by it; in other words, that some bodies have the property of storing light in a state of persisting activity. (Niepce de Saint-Victor, 1861, pp. 44-45)

After many other experiments that cannot be described here, in his fifth communication (1st July 1861), Niepce de Saint-Victor concluded:

> From the whole of my experiments, it follows that this persistent *activity* given by light to all porous bodies, even the most inert ones, cannot be a phosphorescence, because it would not last so long a time, according to the experiments of Mr. Edmond Becquerel. It is therefore more likely that it is a radiation invisible to our eyes, as Mr. Léon Foucault[19] believes, a radiation that does not pass through glass. (Niepce de Saint-Victor, 1861, p. 59)

It was odd, of course, that some kind of radiation that could produce photographic effects could not pass through glass. But the same occurred in the case of ultraviolet light, and Niepce de Saint-Victor also remarked that the light emitted by the combustion of phosphorus in air did not produce photographic effects after passing through glass (Niepce de Saint-Victor, 1859b). At that time, there seemed to be no impossibility in the phenomenon, and it seemed a relevant discovery.

6. REACTIONS TO NIEPCE'S DISCOVERY

Niepce de Saint-Victor's researches were well accepted by the Paris Academy of Science. In 1861 he was unanimously declared the winner of the *Prix Trémont* for his works on photography and light (Chevreul, 1861). The report of the prize committee specially praised the above described experiments:

---

[19] I have not been able to find any paper written by Foucault discussing Niepce's work.

> Mr. Niepce has proved the remarkable fact that some bodies receive from the rays of the Sun the faculty of acting afterwards in darkness upon substances that are sensitive to light, as if those bodies were luminous, in such a way that the Sun transmits to them an activity that they keep for months in darkness. (Chevreul, 1861, p. 1140)

The committee was so strongly impressed by Niepce de Saint-Victor's work that it was suggested that the Trémont Prize for 1862 and 1863 should also be given to him. The Academy accepted the exceptional suggestion. This is a strong evidence that his experiments were accepted as relevant and correct by the French scientific community of that time.

In France, François-Napoléon-Marie Moigno, editor of the journal *Cosmos*, gave ample publicity to the work of his friend Niepce de Saint-Victor (Moigno, 1858). Moigno reproduced in his journal several of Niepce's communications to the Academy of Sciences (Niepce de Saint-Victor 1857b; 1858a; 1859a; 1859c) and presented excited reports such as this:

> That is where we have arrived: we can collect light at the end of the world, carry it to any place we want, and bestow it to our great-nephews, who will be able to use it to reproduce our portrait painted by ourselves or by our own face. Wonder! Wonder! (Moigno, 1858a, p. 286)

Moigno was very well informed about the subject. He published several letters and articles upon Niepce's work, and usually added his own comments. For instance: he reproduced a letter written by a gentleman called Charles Piallat who described some facts that seemed to him similar to those observed by Niepce, but without previous excitation by sunlight. Moigno pointed out that the observed effect was a "Moser image" and was not related to Niepce's invisible phosphorescence (Piallat, 1857).[20]

---

[20] Ludwig Moser's work will be discussed below.

One of the aspects of Niepce's work that called the public attention was the application of his discovery to photography. The time when Niepce de Saint-Victor published his works was a period of fast development of photographic techniques. The use of nitrate of uranium in photographic paper or plates was soon discussed and adopted by many photographers (Hagen, 1859; Blanchère, 1858; Moigno, 1858b; Brébisson, 1858). It seemed to have several advantages: decrease of exposure time, direct acquisition of positive images, etc. It is curious to remark that Niepce was accused of plagiarism by Burnett (1860), who had already used uranium in photography.

In England, Niepce de Saint-Victor's work was also well received. William Grove, one of the early proponents of the "correlation of forces", described Niepce's work to the Royal Institution (Grove, 1858a). He presented it as evidence for his own views on the nature of light (Grove, 1858b). Robert Hunt, a photographic expert, stated that "The recently discoveries of M. Niépce de St. Victor are certainly the most important which have been made since the discovery of photography itself" (Hunt, 1858, p. 15).

Partial translations and comments on Niepce de Saint-Victor's two early papers on "a new action of light" were published by William Crookes, who was the editor of the *Journal of the Photographic Society of London* (Niepce de Saint-Victor, 1857a, 1858c).[21] In editorial announcements of Niepce's works, William Crookes called the attention of the readers, with words of praise:

> We must also direct especial attention to the marvelous discoveries of M. Nièpce de St. Victor, an account of which will be found in our columns. A boundless field for experimental research is therein opened, and we hope that the

---

[21] Both memoirs were published while William Crookes was the editor of that *Journal*. Crookes then founded a new periodical, where he published a translation of Niepce de Saint-Victor's third communication (Niepce de Saint-Victor, 1858b).

> columns of the Journal will soon show that the path of discovery, so grandly pointed out by M. Nièpce, has been quickly followed up by our home experimentalists. (Crookes, 1857, p. 101)

> It will, doubtless, be remembered that in a recent Number of this Journal we published a memoir by M. Nièpce de St. Victor, revealing, among other highly important scientific facts, the singular property which light possesses of communicating to the bodies which absorb it its chemical action on the salts of silver. [...] The sensation excited by these remarkable discoveries throughout Europe has induced M. Nièpce de St. Victor to continue his researches [...]. (Crookes, 1858, p. 169)

The President of the Photographic Society, Sir Frederick Pollock, was deeply affected by Niepce de Saint-Victor's first paper. In the Annual General Meeting of the Society, a significative part of the President's speech celebrated Niepce's achievement (Pollock, 1858). At the same meeting, Niepce de Saint-Victor was elected a honorary member of the Society.

After a short time, however, Niepce's discovery was challenged. It was suggested that the effects observed by Niepce were due to chemical reactions produced by something emanating from the paper excited by sunlight (Laborde, 1858).[22]

In England, doubts were raised on the very *facts* described by Niepce de Saint-Victor. Pollock, who had received so enthusiastically Niepce's researches, was now skeptical:

> I regret to have to inform you that the hopes I gave you last year have not been realised, and that the experiments of M. Nièpce de St. Victor have not been repeated with success by any English experimentalist. I have heard that Mr.

---

[22] An anonymous Italian author also claimed that Niepce's images could be produced by vapours: *Cosmos. Revue Encyclopédique Hebdomadaire des Progrès des Sciences et de leurs Applications aux Arts et a l'Industrie* **13** (1858), 335-339.

Hardwich[23] and several others have tried the experiment and failed. (Pollock, *apud* Crookes, 1859, p. 277; cf. Moigno, 1859a)

Pollock did not doubt the integrity of Niepce. He conjectured that the failure of those duplications could be due to insufficient light intensity, weak sensibility of the photographic paper or some other unknown circumstance.

In order to establish the reality of the phenomenon he had described, Niepce invited Wheatstone to his laboratory in the Louvre and showed him his experiment of light storage in a tube. The experiment succeeded, and Wheatstone took with him two tubes prepared by Niepce, to reproduce the test in England (Moigno, 1859a). Crookes reproduced Moigno's note in his own journal, and added that he had been surprised to see the picture shown to him by Wheatstone in England (Crookes, 1859a, p. 277). Crookes was not convinced, however, that the effect was produced by light, because Niepce himself had also shown that it was possible to produce similar effects using radiant heat.

Niepce's description of the best procedure for obtaining images using stored light led to that interpretation. He recommended that the iron tube should be heated to a temperature of about 60 to 70 degrees (Celsius) before applying it to the photographic plate (Niepce de Saint-Victor, 1859a).

One week after the publication of Niepce's method, Crookes reported that he had repeated the experiment of light storage *without exposing the paper with tartaric acid to light*, but heating it as recommended by Niepce. The hot tube produced the effect that Niepce ascribed to invisible phosphorescence, although the experiment was performed at night and all objects had been kept in darkness before the test. Crookes concluded: "[...] we think we are justified in expressing our opinion that this

---

[23] At that time, Hardwich had not published his experiments. A short report appeared afterwards (Hardwich, 1859).

heat, combined it may be with a chemical reaction between the bodies in the tin tube, is the actual producing cause of the effect he [Niepce] has described" (Crookes, 1859, p. 301)

Moigno defended Niepce:

> It is evident that Mr. Crookes wrongly interpreted his experiment and erroneously concluded that in Mr. Niepce's tubes it is not light that acts. What does this experiment prove, after all? That heat produces in Mr. Crookes' tube the effects that Mr. Niepce ascribes to light. Nothing more, nothing less. (Moigno, 1859b, p. 272)

Moigno's defense was, of course, very weak. Niepce himself provided a stronger answer: he presented new experiments (Niepce de Saint-Victor, 1859c). He used a sheet of paper without tartaric acid or uranium nitrate. He cut it in two pieces, then he kept one of them in darkness and put the other in sunlight. Both pieces were next put into iron tubes. They were afterwards placed upon a photographic plate, in a dark and cold place, for 24 hours. After that time, the photographic plate showed an image produced by the tube containing the paper that received light, and no image at the place where the other tube was applied. Hence, the effect seemed due to light, not to heat.

Niepce also commented that when tartaric acid or uranium salt is added to the paper, the effect is stronger and that it can be observed even when the tubes are not hot. He ascribed to heat a faster release of the stored light and remarked that the tube should not be heated to 100 degrees, to avoid the effect of radiant heat. He also stated that an alkaline photographic paper with silver salt is insensible to heat, and in this way it was possible to distinguish the effects of light and heat.

In France, too, Niepce was attacked. Gaultier de Claubry (1859) was able to reproduce some of his experiments using hot paper as the source of radiation. The temperature used in those experiments, however, was between 100 and 120 degrees – a temperature that should be avoided, according to Niepce.

Bouillon and Sauvage observed that Niepce's tubes produced a faster effect when they were moist and heated than in the case when they were dry heated. They observed that water vapour alone was also able to produce photographic effects. They also reported that Paul Thénard had been able to produce similar effects with a tube containing a paper sheet impregnated by ozone (Bouillon & Sauvage, 1859).

Niepce de Saint-Victor replied by a new experiment: he placed the iron tube containing cardboard with tartaric acid in an ice-box during 48 hours, and even in that case the tube was able to affect photographic paper (Niepce de Saint-Victor, 1859b).

The Abbot Edme César Laborde presented strong evidence for the action of vapours in Niepce's experiments. Niepce's "stored light" did not act through glass. Laborde showed that when a glass plate is put between the tube but at a distance from the photographic plate, allowing circulation of vapours, the plate was affected (Laborde, 1859). He concluded that the effect was produced by formic acid produced by oxidation of the paper used in the experiments. Another author, T. A. Malone, claimed that the effect observed in Niepce's experiments was due to the ink used by French newspapers (Malone, 1860) – an obviously inadequate explanation, since several of Niepce's experiments were made without any printed paper.

No consensus about the explanation of Niepce's phenomenon was reached. After a few years, the whole subject was simply forgotten. After his death, Niepce de Saint-Victor was reminded for his contributions to photographic technique (Larousse, 1865-76, vol. 11, pp. 999-1000; Dreyfus, 1886-1902, vol. 24, p. 1080). However, his researches on invisible phosphorescence sank into oblivion.

Nowadays it is very difficult to understand what happened in Niepce de Saint-Victor's experiments. No single explanation proposed at that time seems to satisfy all observed facts. It is possible that part of the effect he observed was due to the radioactivity of uranium – this would explain the long persistence of the effects and the lack of action through thick glass. On the other hand, according to Niepce, tartaric acid produced effects similar to those of uranium nitrate in several experiments (for instance, the storage experiments), and the experiments seemed to show that heat and excitation by light increased the emission of the invisible radiation by the uranium salt. If part of the effect observed was really due to the radioactivity of uranium, the situation is similar to what happened to Becquerel: the later also stated that another substance (calcium sulphide) emitted penetrating radiations of the same kind as the uranium compounds; and ascribed to the uranium radiation many properties that have not been confirmed afterwards.

7. WAS NIEPCE ANTICIPATED BY MOSER?

Niepce's work was sometimes compared to experiments on invisible light that had been made fifteen years earlier, by Ludwig Moser (see Hunt, 1858; Pollock, 1858, p. 157). A short description of Moser's researches is useful, as it provides another example of the difficulties of understanding an obscure phenomenon.

The period following the invention of the Daguerre photographic process was full of ingenious investigations concerning the effect of light on matter. In a series of papers, Ludwig Moser compared the effects of light, pressure and vapour condensation on several material surfaces. The three kinds of influence were able to produce latent images upon all tested surfaces (Moser, 1842a; Moser, 1843a). According to Moser, if some region of a surface is touched, acted by light or simply breathed on, it acquires the property of precipitating all vapours, which adhere to it (or combine chemically with it) on these spots differently to what it does on the other regions. In this way, any of those influences can produce latent images that can be revealed and fixed by reaction with suitable vapours.

> By these experiments I think I have proved that *contact, condensation of vapours, and light produce the same effect on all bodies.* The differences which appear may be referred to the varying intensity of the producing cause, and the greater or less depth to which the action extends. [...] The most general axiom that I can propose with reference to the influence of the above-mentioned causes is, *that by their means the affinity of all bodies for vapours is modified*, so that they are precipitated and adhere to them in a greater or less degree. (Moser, 1842a; Moser, 1843a, p. 456)

At the end of Moser's first paper there is an *Addendum* where he remarked that contact was not necessary to produce an action upon a surface: one body can act upon another even at a distance, in darkness. This is one of his experiments:

> A plate of agate with several engraved figures was covered with thin strips of mica, and upon these the silver plate was laid, so that the space between the two surfaces amounted to one-fifth of a line, and admitted of seeing through; when, after the lapse of several hours, the plate was introduced into the mercurial vapours, a perfect image of the engraved figures was produced. I have examined other bodies placed at a greater but not measured distance, and always found them depicted, and have thereby discovered the curious fact, that *when two bodies are sufficiently approximated they reciprocally depict each other*. (Moser, 1843a, p. 459)

How did Moser interpret the observed phenomenon? He did not ascribe it to vapours, light or heat. His opinion was '*that every body must be considered as self-luminous*' with emission of an invisible radiation. The phenomenon seemed independent of heat and different from phosphorescence, since, according to Moser, "it made no difference whether the bodies have been kept in the dark for a long time or exposed to the sun before the experiments are made" (Moser, 1843a, 459-460).

In his first paper, Moser described that the following substances were observed to produce effect at a distance upon a silver plate: silver, iodized silver, brass, iron, steel, violet and red glass, black polished horn, white paper, gypsum, mica, agate and cork. In a second paper (Moser, 1842b; Moser, 1843b), written one month later, he added the following to the list of active bodies: gold, copper, German silver, zinc, white transparent glass, bismuth, antimony, tin, lead, mirror metal, wood, mother-of-pearl, black pasteboard, black leather, black velvet, and lamp-black. He generalized his finding to all substances and stated that "it would consequently be a discovery if we could find any body which does not possess self-luminosity, or in which it is present in so small a degree as to escape our observation". Moser also varied the plate upon which the effect was observed, and reported that gold, silver, copper, brass, iron, steel, zinc, porcelain, mica and even liquid mercury were acted by nearby bodies and images could be detected by vapour condensation (Moser, 1843b, pp. 462-463). The images were sharper when the object was closer to the plate receiving the impression. It was possible to obtain images of black characters written on white paper, although Moser remarked that they were not very sharp.

In later papers, Moser published new results. Among other things, he claimed that invisible light (of different colours) was emitted when change of state (condensation of vapour) occurred. He compared the phenomenon to emission of latent heat in changes of state, and hence called that radiation "latent light" (Moser, 1842c; Moser, 1843c).

Moser's work called much attention during a short period. The phenomenon he described was confirmed, but his hypothesis was not accepted. Different explanations of Moser's images were offered by several authors. Erwin Waidele ascribed all effects to exchanges of vapours and gases between sensitive surface and acting body (Waidele, 1845). Hyppolyte-Louis Fizeau explained the images as due to exchanges of greasy organic matter (Fizeau, 1843) and he showed that images were not formed when a thin mica plate was put between the body and the polished plate. M. Knorr and Robert Hunt ascribed the effect to heat (Knorr, 1843; Hunt, 1843). The later showed that the effect occurred even when the surfaces were boiled to eliminate any layer of volatile organic matter. There was no agreement about the agent that produced Moser's images, but there was a consensus that Moser was wrong:

> The beneficial thread that will some day guide us in the tortuous maze of photography has not yet been established; instead of theories we only have more or less plausible hypotheses. We have believed for an instant that Mr. Moser would raise the veil; but he lost himself in the invisible light, as he moved forward guided only by touch [...]. (Moigno, 1847)

Moser images were soon forgotten – the same fate of Niepce's later work.

There are some similarities and many differences between the phenomena described by Moser and Niepce. Moser images were claimed to be produced by any substance upon any other body, without previous excitation by light or heating. In the case of Niepce's researches, some substances were described as able to produce invisible phosphorescence, other substances were apparently inactive; and two substances, in particular, were more active than any other tested by him. Both Moser and Niepce talked about invisible light, but in the first case the emission is supposed to be continuous and spontaneous; on the other hand, the emission of Niepce's invisible light seemed to depend on previous storage of light and could be called, therefore, an invisible phosphorescence.

8. NIEPCE OR BECQUEREL?

Is there any relation between Niepce's invisible light and Henri Becquerel's experiments? Some of Becquerel's contemporaneous scientists thought so. Several scientists recalled Niepce's investigations, after the publication of

Becquerel's researches. One of them was William Crookes. In 1910, he declared:

> Niépce de St. Victor had discovered that Uranium salts possessed the property of storing up light and giving it out in the dark, and in 1858 I took what was perhaps the first radium photograph in this country, by writing with solution of uranium nitrate on a card, insolating it, and then putting it face to face in the dark with a sheet of photographic paper; the image of the writing was reproduced on the paper. (Crookes, 1910, p. 252)[24]

The first mention of Niepce de Saint-Victor's experiments, shortly after Becquerel's experiments, was made in 1896 by Silvanus Thompson:

> It should not be forgotten that so far back as 1857 M. Nièpce de Saint Victor observed many cases in which an object, an engraving on paper or a figured piece of porcelain or marble, immediately after exposure to sunlight, was found capable of giving a photographic impression to a sheet of paper prepared with chloride of silver, with which it was placed in contact. He even used, after exposure to light, cardboard imbibed with salts of uranium or with tartaric acid, and found such to be capable of emitting rays that were photographically active. There was no attempt made, however, to investigate the possibility of transmitting these invisible radiations through opaque bodies. (Thompson, 1896d)

In the same way, Charles Guillaume, in 1897, compared Becquerel's work to that of Niepce de Saint-Victor:

> It is interesting to recall the old experiments of Niepce de Saint-Victor on the emission of radiations in darkness by a

---

[24] It is doubtful that Crookes did that experiment: he never reported it during the period when Niepce's researches were widely discussed.

large number of substances, and particularly by tartaric acid and uranium nitrate. [...] Niepce de Saint-Victor found that those radiations did not traverse glass; but this result could be due to a lack of sensitivity of his plates. (Guillaume, 1897, p. 133, footnote)

One of the critics of Becquerel, that reminded him several times of Niepce de Saint-Victor's former work, was Gustave le Bon, who explicitly ascribed to Niepce the discovery of the emission of uranium rays (Le Bon, 1900, p. 299, footnote 1). In a later publication, Le Bon even accuses Becquerel of appropriation of Niepce's ideas without due acknowledgment (Le Bon, 1907, pp. 21-22, 424-426).

An anonymous (and not well informed) review published in December 1897 described Le Bon's "black light" experiments as repetition of Niepce de Saint-Victor's observations (Anonymous, 1897). The author recalled that one argument against Niepce de Saint-Victor's claims was the lack of action of the tubes containing uranium nitrate when a glass was interposed between the tube and the photographic plate; however, according to this author, this was not as conclusive as was thought, because there do exist radiations that would be absorbed by thick glass.

In his 1903 book, one finds the only comments Becquerel ever made on Niepce de Saint-Victor's work.

> When I published my first observations about uranium radiation, some people have tried to bring together those statements to experiences formerly made by Mr. Niepce de Saint-Victor with several papers, some of which were impregnated with tartaric acid or uranium nitrate [...].
>
> Although Foucault had issued the hypothesis of an unknown radiation to explain those phenomena, it was proved that this effect, that was not produced through glass or through a thin slab of mica, was due to chemical actions arising from the decomposition of organic or saline matter by light. It is true that uranium nitrate is among those substances. [...] On those papers, uranium is in such a small amount that in order

> to produce an appreciable impression on the photographic
> plates used by the author, several months of exposition would
> have been necessary. Therefore, Mr. Niepce de Saint-Victor
> has not been able to observe uranium radiation. (Becquerel,
> 1903a, pp. 51-52)

Becquerel added that the impossibility of reproducing Niepce de Saint-Victor's effects with mica or glass between the photographic paper and uranium proves that the observed phenomena were not produced by uranium rays. He also states that he tried to reproduce Niepce de Saint-Victor's experiments with tartaric acid and plain paper, with black paper between the active substance and a photographic plate or an electroscope, and observed no effect.

Was Henri Becquerel aware of Niepce de Saint-Victor's work when he began his studies of "radioactivity"? Probably not, because that work had long been forgotten and was not cited by his father.

In 1867, Edmond Becquerel published a monumental work on phosphorescence: *La lumière, ses causes et ses effects*. After publication of this book, no other comprehensive treatise on this subject appeared in any language until the next century (Harvey, 1957, p. 221). The effect observed by Niepce de Saint-Victor was not described in Edmond Becquerel's book.[25] It was not, however, completely neglected by other contemporary phosphorescence researchers. It was referred in a book published by Thomas Lamb Phipson (1870, pp. 72-76), who described the "invisible phosphoresce" discovered by his "ingenious friend" Niepce de Saint-Victor. Phipson reported that he had seen experiments with tubes containing cardboard imbided with tartaric acid or uranium salt, that produced photographic effects a few months after their exposure to light. Phipson did not describe any criticism concerning Niepce's

---

[25] In this book, there is a passing mention of the use of uranium salts in photography, but no direct reference to Niepce's work (Edmond Becquerel, 1867-1868, vol. 2, 73).

experiments. If Henri Becquerel had read Phipson's book, that chapter could have suggested him the idea of invisible phosphorescence of uranium compounds. There is no evidence, however, that he ever read the book.

9. THE CONCEPT OF SCIENTIFIC DISCOVERY

The process of discovery is a very complex process. The following analysis is an intended contribution to the elucidation of the meaning of scientific discovery of new phenomena[26]. "Discovery" is not a technical term with well defined meaning. For that reason, although the analysis presented below conforms to some current uses of "discovery", it cannot conform to all different and mutually contradictory uses of the word.

I will call "discovery of a new scientific phenomenon" a complex operation that includes, among other things, the individual (or group) research work, communication and social acceptance of the following aspects:

1) To have contact with the phenomenon.
> The researcher might meet or to find a new phenomenon by chance or while looking for it. Prediction without actual detection cannot be called a "discovery".

2) To notice that it is a new phenomenon.
> The researcher can only conclude that something is new by comparing it with other known similar phenomena, remarking relevant differences. If a researcher is in contact with a phenomenon, but does not notice that it is different from known phenomena, he made no discovery.

3) To identify the phenomenon.

---

[26] The concept of phenomenon used in this paper agrees in most aspects with Hacking's use of the term (Hacking, 1983, chapter 13, especially p. 225). However, I prefer to use "to discover" than "to invent" a phenomenon, even when it is an effect artificially created in the laboratory.

To identify a phenomenon is different from understanding it. In order to allow further investigation of some phenomenon, it is necessary to recognize when it is present and when it is not, noticing similarities and distinguishing the phenomenon from similar but different phenomena. The characterization of a new phenomenon is usually dynamic: it changes in time. However, *some* identification (even if temporary) is desirable. When several different phenomena are confounded, the specificity of the new phenomenon is lost.

4) To identify conditions that make the phenomenon and its main effects reproducible.

This step corresponds to the detection of relevant conditions and/or causes of the phenomenon. When this aspect of the investigation is fulfilled, the phenomenon becomes reproducible, if the conditions can be controlled. Of course, further research may show that some conditions that seemed irrelevant are important, and vice-versa; but it is desirable to *attempt* the identification of the relevant conditions.

5) To find and to provide adequate evidence for some properties (generalizations and exceptions) of the phenomenon.

The empirical investigation of the phenomenon provides data that must be analyzed in order to infer the properties of the phenomenon.[27] In this step, there is an interplay of facts and arguments – data and interpretation. The facts should be carefully collected and tested, in order to avoid wrong data; and the arguments should be sound. Whenever possible, quantitative aspects should be measured and empirical laws should be proposed.

6) To understand the phenomenon.

---

[27] Cf. the analysis of *phenomena* and *data* presented by James Woodward and James Bogden (1988).

It is desirable to suggest a plausible interpretation (one that does not directly conflict with known facts), to provide evidence for or against hypotheses, to present evidence to distinguish between alternative interpretations, etc.

7) To include the phenomenon in the domain of a broad scientific theory.

The integration of a new phenomenon to a wider scientific theory is the final *desideratum* of the discovery. The integration presupposes the empirical knowledge of a large number of properties of the phenomenon and the study of compatibility between them and consequences of the theory.

I propose to classify as an *empirical discovery* of a scientific phenomenon the fulfillment of desiderata 1 to 5. The scientific relevance of the phenomenon, however, depends on the fulfillment of desiderata 6 and 7, because an isolated, unexplained phenomenon, is of lower scientific value. We might call the fulfillment of all desiderata 1 to 7 the *complete* or *full discovery* of the phenomenon.

Scientific knowledge is never complete. It is always possible to find out new conditions that influence some phenomenon, or to find some new properties of an old phenomenon. The discovery is accepted as such after *some* relevant conditions and *some* properties of the phenomenon have been established – completeness is neither required nor possible.

Scientific knowledge is temporary. Therefore, at any given time, there might be accepted properties and hypotheses that are afterwards dismissed as wrong. At each time, the researcher who found and provided evidence for the accepted knowledge concerning the phenomenon can be regarded as its discoverer. If, however, at another time, the data and interpretation provided by some researcher are rejected, he will no longer be called the discoverer of the effect – except in ironic phrases, such as "Blondot was the discoverer of N-rays".

According to the above analysis, Röntgen was the *empirical discoverer* of X-rays because he had contact with the phenomenon, noticed that it was a new phenomenon, identified it, established conditions that made the phenomenon and its main effects reproducible, and found and provided adequate evidence for some properties (generalizations and exceptions) of the phenomenon. All the properties of X-rays described by Röntgen were confirmed by later observers. Of course, many properties that were not described by him were found by other researchers, but Röntgen had already shown, in his first paper, the existence of a new, reproducible, recognizable phenomenon. Röntgen was unable, however, to understand the nature of X-rays and to link it to broader physical theories.

Full discoveries are usually preceded by prediction. The full discovery of electromagnetic waves was done by Hertz – the theoretical explanation was already available, of course, but Hertz was able to produce the waves, to study their properties and to exhibit an agreement between theory and experiment.

Discoveries can seldom be ascribed to a single person. Even in the case of empirical discoveries, contributions from different researchers might be necessary before the phenomenon is adequately described. For instance: in the case of the discovery of Brownian motion, there was a wide gap between the early observation and description of motion of microscopic particles in a liquid and the elucidation of the relevant conditions and properties of the phenomenon.

10. CONCLUSION: THE DISCOVERY OF RADIOACTIVITY

After all, can anyone claim that Niepce de Saint-Victor discovered radioactivity? Yes, if we reduce "discovery" to the first contact with a phenomenon. No, if discovery of a phenomenon also implies the discrimination between that phenomenon and other similar but distinct phenomena, the correct ascertainment of properties and the correct interpretation of phenomena. According to the analysis of "discovery" presented above, the empirical discovery of radioactivity cannot be ascribed to Niepce.

Can Henri Becquerel be called the discoverer of radioactivity? No. If we reduce "discovery" to the first contact with a phenomenon, Becquerel was preceded by Niepce. If discovery of a phenomenon also implies the discrimination between that phenomenon and other similar but distinct phenomena, the correct ascertainment of properties and the correct interpretation of phenomena, neither Becquerel nor Niepce discovered it.

Why, then, did the Royal Swedish Academy of Sciences ascribe the discovery of radioactivity to Becquerel? Did the Swedish Academy assume another different criterion for assigning the discovery?

When the Nobel Prize was given to Henri Becquerel, the President of the Swedish Academy presented a justification. According to this presentation speech, Becquerel

> [...] demonstrated that these substances [salts of uranium] emit rays of a special nature, distinct from ordinary light. Tests continued and he established an even more extraordinary fact, namely that this radiation is not in direct relation to the phenomenon of phosphorescence, that phosphorescent as well as those which are not can give rise to this radiation, that previous lighting is never necessary for the phenomenon to occur, and lastly that the radiation in question continues with invariable force to all appearances without its origin being traced to any of the known forces of energy. This was how Becquerel made the discovery of *spontaneous radioactivity* and the rays that bear his name. (Törnebladh, 1903, p. 14)

> Becquerel had already shown by the study of uranium radiation some of the most important properties of those rays.[...]
> Becquerel radiation resembles light in several respects. Propagation is rectilinear. [...] Yet it differs from light in certain essentials, for example by its property to pass through

metals [...] and lastly by the absence of phenomena of reflection, interference and refraction, characteristic of light. In this the Becquerel rays are exactly similar to Röntgen rays and cathode rays. It has been found all the same that Becquerel radiation is not homogeneous, but is composed of different kinds of rays [...] (Törnebladh, 1903, pp. 15-16)

According to the analysis of "discovery" presented above, if Becquerel had done what the President of the Swedish Academy of Sciences described, he would have deserved the epithet of [empirical] discoverer of radioactivity. However, as a matter of historical fact, Becquerel did not do what was ascribed to him. It was not because of a different concept of "discovery" or due to the use of different criteria that the Nobel Prize was given to Becquerel – it was because the Swedish Academy of Sciences was ill informed about the real contribution of Becquerel and other researchers to the knowledge of radioactivity. This misinformation was not due to chance: it was due to Henri Becquerel's systematic propaganda strategy.[28]

As it often occurs, the full discovery radioactivity was the result of a gradual and collective effort. Its beginning was the search for penetrating radiations emitted by luminescent bodies, motivated by Poincaré's conjecture. It was completed by the development of Ernest Rutherford and Frederick Soddy's theory of transmutation of the elements.

ACKNOWLEDGEMENTS

This work was produced while the author was a visiting scholar of the Department of History and Philosophy of Science, University of Cambridge, and visiting fellow of Wolfson College (1995-1996). The author is grateful to the São Paulo State Research Foundation, Brazil (FAPESP) for supporting this research.

---

[28] See MARTINS, Roberto de Andrade. Becquerel's experimental mistakes, in this volume.

BIBLIOGRAPHIC REFERENCES

[ANONYMOUS]. Uranium in photography. *The British Journal of Photography* **44**: 770-771, 1897.

BADASH, Lawrence. "Chance favors the prepared mind": Henri Becquerel and the discovery of radioactivity. *Archives Internationales d'Histoire des Sciences* **18** (70): 55-66, 1965 (a).

BADASH, Lawrence. Radioactivity before the Curies. *American Journal of Physics* **33**: 128-135, 1965 (b).

BADASH, Lawrence. Becquerel's "unexposed" photographic plates. *Isis* **57**: 267-269, 1966.

BECQUEREL, Edmond. Recherches sur divers effets lumineux qui résultent de l'action de la lumière sur les corps. *Annales de Chimie et de Physique* [3] **55**: 5-119, 1859.

BECQUEREL, Edmond. *La lumière, ses causes et ses effets.* 2 vols. Paris: Firmin Didot, 1867-1868.

BECQUEREL, Henri. Sur les radiations émises par phosphorescence. *Comptes Rendus Hebdomadaires des Séances de l'Académie des Sciences de Paris* **122**: 420-421, 1896 (a).

BECQUEREL, Henri. Sur les radiations invisibles émises par les corps phosphorescents. *Comptes Rendus Hebdomadaires des Séances de l'Académie des Sciences de Paris* **122**: 501-503, 1896 (b).

BECQUEREL, Henri. Sur quelquer propriétés nouvelles des radiations invisibles émises par divers corps phosphorescents. *Comptes Rendus Hebdomadaires des Séances de l'Académie des Sciences de Paris* **122**: 559-564, 1896 (c).

BECQUEREL, Henri. Sur les radiations invisibles émises par les sels d'uranium. *Comptes Rendus Hebdomadaires des Séances de l'Académie des Sciences de Paris* **122**: 689-694, 1896 (d).

BECQUEREL, Henri. Sur les propriétés différentes des radiations invisibles émises par les sels d'uranium, et du rayonnement de la paroi anticathodique d'un tube de

Crookes. *Comptes Rendus Hebdomadaires des Séances de l'Académie des Sciences de Paris* **122**: 762-767, 1896 (e).

BECQUEREL, Henri. Émission de radiations nouvelles par l'uranium métallique. *Comptes Rendus Hebdomadaires des Séances de l'Académie des Sciences de Paris* **122**: 1086-1088, 1896 (f).

BECQUEREL, Henri. Sur diverses propriétés des rayons uraniques. *Comptes Rendus Hebdomadaires des Séances de l'Académie des Sciences de Paris* **123**: 855-858, 1896 (g).

BECQUEREL, Henri. Recherches sur les rayons uraniques. *Comptes Rendus Hebdomadaires des Séances de l'Académie des Sciences de Paris* **124**: 438-444, 1897 (a).

BECQUEREL, Henri. Sur la loi de la décharge dans l'air de l'uranium électrisé. *Comptes Rendus Hebdomadaires des Séances de l'Académie des Sciences de Paris* **124**: 800-803, 1897 (b).

BECQUEREL, Henri. Note sur quelques propriétés du rayonnement de l'uranium et des corps radio-actifs. *Comptes Rendus Hebdomadaires des Séances de l'Académie des Sciences de Paris* **128**: 771-777, 1899.

BECQUEREL, Henri. Recherches sur une propriété nouvelle de la matière – activité radiante spontanée ou radioactivité de la matière. *Mémoires de l'Académie des Sciences de l'Institut de France* **46**: 1-360, 1903 (a).

BECQUEREL, Henri. *Recherches sur une propriété nouvelle de la matière: activité radiante spontanée ou radioactivité de la matière*. Paris: Firmin-Didot, 1903 (b).

BECQUEREL, Jean, *La radioactivité et les transformations des éléments*. Paris: Payot, 1924.

BERTRAND, Gabriel. Sur l'origine de la découverte de la radioactivité. *Comptes Rendus Hebdomadaires des Séances de l'Académie des Sciences de Paris* **223**: 698-670, 1946.

BLANCHÈRE, M. de la. Méthode operatoire pour obtenir les épreuves positives avec les sels d'urane d'après la découverte de M. Niepce de Saint-Victor. *Cosmos. Revue Encyclopédique Hebdomadaire des Progrès des Sciences et de leurs Applications aus Arts e a l'Industrie* **12**: 292-296, 398-402, 1858.

BOUILLON, E. & SAUVAGE, A. Photographie et thermographie. *Cosmos. Revue Encyclopédique Hebdomadaire des Progrès des Sciences et de leurs Applications aus Arts e a l'Industrie* **14**: 511-514, 1859.

BRÉBISSON. Positifs à l'azotate d'urane. *Cosmos. Revue Encyclopédique Hebdomadaire des Progrès des Sciences et de leurs Applications aus Arts e a l'Industrie* **13**: 395-396, 1858.

BURNETT, C. J. [Letter] To the editor of the Photographic Journal. *Journal of the Photographic Society of London* **6**: 5, 1860.

CHEVREUL, Michel Eugène. Rapport sur le concours pour le Prix Trémont, année 1861. *Comptes Rendus Hebdomadaires des Séances de l'Académie des Sciences de Paris* **53**: 1139-1140, 1861.

COLSON, René. Rôle des différentes formes de l'énergie dans la photographie au travers des corps opaques. *Comptes Rendus Hebdomadaires des Séances de l'Académie des Sciences de Paris* **122**: 598-600, 1896.

CROOKES, William. [Editorial. No. 61. December 21, 1857]. *Journal of the Photographic Society of London* **4**: 101, 1857.

CROOKES, William. [Editorial. No. 64. March 22, 1858]. *Journal of the Photographic Society of London* **4**: 169, 1858.

CROOKES, William. Ocular demonstration of M. Nièpce's discovery with respect to light. *The Photographic News* **1**: 277-278, 1859 (a).

CROOKES, William. Photographs in the dark. Observations on M. Nièpce's discovery of "A new action of light". *The Photographic News* **1**: 301, 1859 (b).

CROOKES, William. The production of photographs without the aid of light. *The Photographic News* **2**: 17, 1859 (c).

CROOKES, William. [Speech at a banquet to past presidents of the Chemical Society]. *Chemical News* **102**: 252-253, 1910.

CURIE, Marie Sklodowska. Rayons émis par les composés de l'uranium et du thorium. *Comptes Rendus Hebdomadaires des Séances de l'Académie des Sciences de Paris* **126**: 1101-1103, 1898.

CURIE, Pierre & CURIE, Marie Sklodowska. Les nouvelles substances radioactives et les rayons qu'elles émettent. Vol. 3, pp. 79-113, *in*: GUILLAUME, Charles-Édouard & POINCARÉ, Lucien (eds.). *Rapports presentes au Congrès International de Physique réuni a Paris en 1900*. 3 vols. Paris: Gauthier-Villars, 1900.

DREYFUS, Camille (ed.). *La Grande Encyclopédie. Inventoire Raisonné des Sciences, des Lettres et des Arts*. 31 vols. Paris: Société Anonyme de "La Grande Encyclopédie", 1886-1902.

FIZEAU, Hyppolyte-Louis. Sur les causes qui concourent à la production des images de Moser. *Annales de Chimie et de Physique* [3] **7**: 240-241, 1843.

GAULTIER DE CLAUBRY, Henri François. Action de la chaleur dans la formation de certaines images photographiques. *Comptes Rendus Hebdomadaires des Séances de l'Académie des Sciences de Paris* **48**: 811, 1859.

GIBSON, Charles R. The history of photography. Pp. 1-34, *in*: CONRADY, Alexander Eugen (ed.). *Photography as a scientific implement*. London: Blackie and Son, 1923.

GROVE, William Robert. On molecular impressions by light and electricity. *Notices of the Proceedings at the Meetings of the Members of the Royal Institution of Great Britain* **2**: 458-464, 1858.

GUILLAUME, Charles-Édouard. *Les rayons X et la photographie a travers les corps opaques*. 2nd edition. Paris: Gauthier-Villars et Fils, 1897.

HACKING, Ian. *Representing and intervening*. Cambridge: Cambridge University Press, 1983.

HAGEN, O. On the employment of nitrate of uranium in photography. *Journal of the Photographic Society of London* **5**: 75-76, 1858.

HARDWICH, F. M. Nièpce's "New action of light". *The Photographic News* **2**: 46, 1859.

HARVEY, E. Newton. *A history of luminescence from the earliest times until 1900*. Philadelphia: The American Philosophical Society, 1957.

HENRY, Charles. Augmentation du rendement photographique des rayons Roentgen par le sulfure de zinc phosphorescent. *Comptes Rendus Hebdomadaires des Séances de l'Académie des Sciences de Paris* **122**: 312-314, 1896.

HUNT, Robert. On the spectral images of M. Moser; a reply to his animadversions, &c. *London, Edinburgh and Dublin Philosophical Magazine and Journal of Science* [3] **23**: 415-416, 1843.

HUNT, Robert. Niepce de Saint Victor's discovery of new and remarkable photographic phenomena. *The Art Journal* [2] **4**: 15-16, 1858.

JAUNCEY, George Eric Macdonnell. The birth and early infancy of X-rays. *American Journal of Physics* **13**: 362-379, 1945.

KNORR, M. Sur la formation des images de Moser. *Annales de Chimie et de Physique* [3] **7**: 239-240, 1843.

LABORDE, Abbé Edme César. Photographie. Conservation des papiers positifs. *Cosmos. Revue Encyclopédique Hebdomadaire des Progrès des Sciences et de leurs Applications aus Arts e a l'Industrie* **13**: 149-150, 1858.

LABORDE, Abbé Edme César. Photographie. Sur l'activité persistante de la lumière. *Cosmos. Revue Encyclopédique Hebdomadaire des Progrès des Sciences et de leurs Applications aus Arts e a l'Industrie* **15**: 266-268, 1859.

LAROUSSE, Pierre (ed.). *Grand dictionnaire universel.* 15 vols. Paris: Administration du Grand Dictionnaire, 1865-1876.

LE BON, Gustave. Les formes diverses de la phosphorescence. *Revue Scientifique* [4] **14**: 289-305, 1900.

LE BON, Gustave. *L'évolution de la matière*. Paris: Flammarion, 1906.

LE BON, Gustave. *The evolution of matter*. Translated by F. Legge. London: The Walter Scott Publishing Co, 1907.

MALONE, T. A. On an alleged new property of light. *The Photographic News* **4**: 329, 1860.

MALONE, T. A. On an alleged new property of light. *Journal of the Photographic Society of London* **7**: 38-39, 1862.

MAY, Kenneth O. Historiographic vices. II. Priority chasing. *Historia Mathematica* **2**: 315-317, 1975.

MENTRÉ, François. Histoire des sciences. L'attribution et le baptême des découvertes. *Revue Scientifique* [5] **4**: 489-495, 1905.

MOIGNO, François Napoléon Marie. Photographie, son histoire, ses procédés, sa théorie. *Revue Scientifique et Industrielle* [2] **14**: 231-280, 321-374, 1847.

MOIGNO, François Napoléon Marie. Nouvelles de la semaine. *Cosmos. Revue Encyclopédique Hebdomadaire des Progrès des Sciences et de leurs Applications aus Arts e a l'Industrie* **12**: 281-286, 1858 (a).

MOIGNO, François Napoléon Marie. Procédé de photographie au nitrate d'urane. *Cosmos. Revue Encyclopédique Hebdomadaire des Progrès des Sciences et de leurs Applications aus Arts e a l'Industrie* **13**: 277-281, 1858 (b).

MOIGNO, François Napoléon Marie. Positifs à l'azotate d'urane. *Cosmos. Revue Encyclopédique Hebdomadaire des Progrès des Sciences et de leurs Applications aus Arts e a l'Industrie* **13**: 395, 1858 (c).

MOIGNO, François Napoléon Marie. Activité persistante de la lumière. *Cosmos. Revue Encyclopédique Hebdomadaire des Progrès des Sciences et de leurs Applications aus Arts e a l'Industrie* **14**: 153-155, 1859 (a).

MOIGNO, François Napoléon Marie. Photographie dans l'obscurité. *Cosmos. Revue Encyclopédique Hebdomadaire des Progrès des Sciences et de leurs Applications aus Arts e a l'Industrie* **14**: 272-273, 1859 (b).

MOSER, Ludwig. Ueber den Process des Sehens und die Wirkung des Lichts auf alle Körper. *Annalen der Physik und Chemie* **56**: 177-234, 1842 (a).

MOSER, Ludwig. Einige Bemerkungen über das unsichtbare Licht. *Annalen der Physik und Chemie* **56**: 569-574, 1842 (b).

MOSER, Ludwig. Ueber das Latentwerden des Lichts. *Annalen der Physik und Chemie* **57**: 1-34, 1842 (c).

MOSER, Ludwig. On vision, and the action of light on all bodies. Translated by Henry Croft. *Scientific Memoirs Selected from the Transactions of Foreign Academies of Science and Learned Societies and from Foreign Journals, edited by Richard Taylor* **3**: 422-461, 1843 (a).

MOSER, Ludwig. On vision, and the action of light on all bodies. Translated by Henry Croft. *Scientific Memoirs Selected from the Transactions of Foreign Academies of Science and Learned Societies and from Foreign Journals, edited by Richard Taylor* **3**: 461-464, 1843 (b).

MOSER, Ludwig. On vision, and the action of light on all bodies. Translated by Henry Croft. *Scientific Memoirs Selected from the Transactions of Foreign Academies of Science and Learned Societies and from Foreign Journals, edited by Richard Taylor* **3**: 465-487, 1843 (c).

NIEPCE DE SAINT-VICTOR, Abel. Sur les propriétés photogéniques de l'iode, du phosphore, de l'acide azotique, &c. *Comptes Rendus Hebdomadaires des Séances de l'Académie des Sciences de Paris* **25**: 579-589, 1847.

NIEPCE DE SAINT-VICTOR, Abel. Sur les propriétés photogéniques de l'iode, du phosphore, de l'acide azotique, &c. *Annales de Chimie* **22**: 85-97, 1848; *Philosophical Magazine* **32**: 206-215, 1848 (a).

NIEPCE DE SAINT-VICTOR, Abel. Note sur la photographie sur verre. *Comptes Rendus Hebdomadaires des Séances de l'Académie des Sciences de Paris* **26**: 637-639, 1848 (b).

NIEPCE DE SAINT VICTOR, Abel. Sur une nouvelle action de la lumière. *Comptes Rendus Hebdomadaires des Séances de l'Académie des Sciences de Paris* **45**: 811-815, 1857; **46**: 448-452, 489-491, 1858; **47**: 866-869, 1002-1006, 1858; **53**: 33-35, 1861; **65**: 505-506, 1867.

NIEPCE DE SAINT-VICTOR, Abel. Note sur des images du Soleil et de la Lune obtenues par la photographie sur verre. *Comptes Rendus Hebdomadaires des Séances de l'Académie des Sciences de Paris* **30**: 709-11, 1850 (a).

NIEPCE DE SAINT-VICTOR, Abel. Note sur la photographie sur verre et sur quelques faits nouveaux. *Comptes Rendus Hebdomadaires des Séances de l'Académie des Sciences de Paris* **31**: 245-7, 1850 (b).

NIEPCE DE SAINT-VICTOR, Abel. Memoir on a new action of light. *Journal of the Photographic Society of London* **4**: 122-124, 1857 (a).

NIEPCE DE SAINT-VICTOR, Abel. Phosphorescence et fluorescence mises en évidence au moyen de la photographie. *Cosmos. Revue Encyclopédique Hebdomadaire des Progrès des Sciences et de leurs Applications aus Arts e a l'Industrie* **11**: 567-571, 1857 (b).

NIEPCE DE SAINT-VICTOR, Abel. Deuxième mémoire sur une nouvelle action de la lumière. *Cosmos. Revue Encyclopédique Hebdomadaire des Progrès des Sciences et de leurs Applications aus Arts e a l'Industrie* **12**: 268-274, 1858 (a).

NIEPCE DE SAINT-VICTOR, Abel. On an action of light hitherto unknown. *The Photographic News* **1**: 158-159, 194-195, 1858 (b).

NIEPCE DE SAINT-VICTOR, Abel. Second memoir on a new action of light. *Journal of the Photographic Society of London* **4**: 179-182, 1858 (c).

NIEPCE DE SAINT-VICTOR, Abel. Activité persistante de la lumière. *Cosmos. Revue Encyclopédique Hebdomadaire des Progrès des Sciences et de leurs Applications aux Arts et a l'Industrie* **14**:183-184, 1859; translated in: The Photographic News **1**: 291, 1859 (a).

NIEPCE DE SAINT-VICTOR, Abel. Note sur l'activité communiquée par la lumière au corps qui a été frappé par elle. *Comptes Rendus Hebdomadaires des Séances de l'Académie des Sciences de Paris* **48**: 741-742, 1859. English translation: 'Preserved light', *The Photographic News* **2**: 87-88, 1859 (b).

NIEPCE DE SAINT-VICTOR, Abel. Note sur l'activité communiquée par la lumière au corps qui a été frappé par elle. *Cosmos. Revue Encyclopédique Hebdomadaire des Progrès des Sciences et de leurs Applications aux Arts et a l'Industrie* **14**: 389-390, 1859. English translation: Abel Niepce de Saint-Victor, 'The new action of light', *The Photographic News* **2:** 62, 1859 (c)

NIEPCE DE SAINT-VICTOR, Abel. Sur une nouvelle action de la lumière. *Mémoires de l'Académie Impériale des Sciences, Arts et Belles-Lettres de Dijon. Section des Sciences* [2] **9**: 33-59, 1861.

NIEPCE DE SAINT-VICTOR, Abel. Sur une nouvelle action de la lumière. *Comptes Rendus Hebdomadaires des Séances de l'Académie des Sciences de Paris* **65**: 505-7, 1867.

NIEWENGLOWSKI, Gaston Henri. Sur la proprieté qu'ont les radiations émises par les corps phosphorescents, de traverser certains corps opaques à la lumière solaire, et sur les expériences de M. G. Le Bon, sur la lumière noire. *Comptes Rendus Hebdomadaires des Séances de l'Académie des Sciences de Paris* **122**: 385-386, 1896.

PHIPSON, Thomas Lamb. *Phosphorescence, or the emission of light by minerals, plants and animals*. 2nd. ed. London: L. Reeve, 1870.

PIALLAT, Charles. Photographie. *Cosmos. Revue Encyclopédique Hebdomadaire des Progrès des Sciences et de leurs Applications aus Arts e a l'Industrie* **11**: 686-687, 1857.

POINCARÉ, Henri. Les rayons cathodiques et les rayons Roentgen. *Revue Générale des Sciences* **7**: 52-59, 1896.

POINCARÉ, Henri. Les rayons cathodiques et les rayons Roentgen. *Revue Scientifique* [4] **7**: 72-81, 1897.

POLLOCK, Frederick. Photographic Society. Annual general meeting. Tuesday, 2nd. February 1858 [Presidential address]. *Journal of the Photographic Society of London* **4**: 154-157, 1858.

POTONNIÉE, Georges. *The history of the discovery of photography*. New York: Tennant and Ward, 1936.

ROMER, Alfred. *The discovery of radioactivity and transmutation*. New York: Dover, 1964.

ROMER, Alfred. *Radiochemistry and the discovery of isotopes*. New York: Dover, 1970.

ROMER, Alfred. Becquerel, [Antoine-] Henri. Vol. 1, pp. 558-561, in: GILLISPIE, Charles C. (ed.). *Dictionary of scientific biography*. New York: Charles Scribner's Sons, 1981.

SAGNAC, Georges. Les expériences de M. H. Becquerel sur les radiations invisibles émises par les corps phosphorescents et par les sels d'uranium. *Journal de Physique Théorique et Appliquée*, [série 3], **5**: 193-202, 1896 (b).

SAMUELSSON, Bergt; SOHLMAN, Michael (eds.). *Nobel lectures including presentation speeches and laureates' biographies*. Physics. 1901-1921. Amsterdam: Elsevier, 1967.

SCHMIDT, Gerhard C. Ueber die von den Thorverbindungen und einigen anderen Substanzen ausgehende Strahlung. *Annalen der Physik und Chemie* [2] **65**: 141-151, 1898. Reprinted in: *Verhandlungen der physikalische Gesellschaft nach Berlin*, **17**: 14-16, 1898.

STEWART, Oscar M. A résumé of the experiments dealing with the properties of Becquerel rays. *Physical Review* **6**: 239-51, 1898.

STEWART, Oscar M. Becquerel rays, a résumé. *Physical Review* **11**: 155-75, 1900.

THOMPSON, Silvanus P. On hyperphosphorescence. *The London, Edinburgh and Dublin Philosophical Magazine and Journal of Science* [série 5] **42**: 103-107, 1896 (d).

THOMPSON, Silvanus P. *Light visible and invisible.* London: MacMillan and Co., 1897.

THOMSON, Joseph John. The Röntgen rays. *Nature* **53**: 581-583, 1896 (c).

TÖRNEBLADH, Henricus Ragnar. Prix Nobel de physique. *Les Prix Nobel* **3**: 13-17, 1903.

WAIDELE, Erwin. Versuche und Beobachtungen über Prof. Moser's unsichtbares Licht. *Annalen der Physik und Chemie* **59**: 255-283, 1845.

WASSON, Tyler (ed.). *Nobel Prize winners.* New York: H. W. Wilson, 1987.

WOODWARD, James & BOGEN, James. Saving the phenomena. *The Philosophical Review* **97**: 303-352, 1988.

BECQUEREL'S EXPERIMENTAL MISTAKES

Roberto de Andrade Martins

Abstract: In 1896, Henri Becquerel detected a penetrating radiation emitted by some uranium salts and met a phenomenon that nowadays we call "radioactivity". Becquerel's study of uranium radiation was not casual or blind. It was guided by his acceptance of Poincaré's conjecture concerning a possible relation between X rays and luminescence. What Becquerel expected to find was the emission of a penetrating electromagnetic radiation (something similar to ultraviolet rays) emitted by a special phenomenon of fluorescence or phosphorescence that violated Stokes' law. Guided by his preconceptions, Becquerel described experiments that seemed to support the view that uranium radiation had the usual properties of known electromagnetic waves: reflection, refraction and polarization. He also described an increase in the emission of radiation whcn uranium compounds wcrc stimulated by sunlight. Those and several other aspects of Becquerel's experimental work must nowadays be interpreted as experimental mistakes. Becquerel's mistakes were gradually corrected by other researchers. As the study of radioactivity developed, Becquerel reinterpreted his own early work, hiding his mistakes or ascribing to himself their correction. The aim of this article is to discuss one particular episode of experimentation – Becquerel's study of the phenomenon we call radioactivity – and the methodological problems aroused by his mistakes.
Keywords: radioactivity; experimental errors; history of physics; Becquerel, Henri

MARTINS, Roberto de Andrade. *Historical Essays on Radioactivity*. Extrema: Quamcumque Editum, 2021.

1. INTRODUCTION

The study of experimental research has recently deserved more attention from historians of science (Holmes, 1992). The aim of this article is to discuss one particular episode of experimentation – Becquerel's study of the phenomenon we call radioactivity – and the methodological problems aroused by his mistakes.

In 1896, Henri Becquerel detected a penetrating radiation emitted by some uranium salts and came accross a phenomenon that nowadays we call "radioactivity". Becquerel's study of uranium radiation was not casual or blind. It was guided by his acceptance of Poincaré's conjecture[1], together with his previous knowledge and expectations concerning the properties of uranium compounds (Martins, 1997). What Becquerel expected to find was the emission of penetrating electromagnetic radiation (something similar to ultraviolet rays) emitted by a special kind of fluorescence or phosphorescence that violated Stokes' law. That was, indeed, what he thought he had found. Guided by his preconceptions, Becquerel ascribed to uranium radiation the usual properties of known electromagnetic waves: reflection, refraction and polarization. As he thought the phenomenon to be a kind of phosphorescence, he also expected to observe an increase in the emission of radiation when uranium compounds were stimulated by sunlight – and he verified this increase.

Those and several other aspects of Becquerel's experimental work must nowadays be interpreted as full of experimental mistakes. There is nothing new in the observation that scientists sometimes are misled by their theoretical expectations – but it is remarkeable how far Becquerel was led by his preconceptions. He was even led to support the claims for existence of N-rays, because they seemed able to explain some of his own experimental anomalies.

---

[1] See the first paper in this volume: MARTINS, Roberto de Andrade. A pool of radiations: Becquerel and Poincaré's conjecture.

Becquerel's mistakes were gradually corrected by other researchers. As the study of radioactivity developed, Becquerel reinterpreted his own early work, hiding his mistakes or ascribing to himself their correction. He was successful, and in a few years his errors were forgotten – and he was accorded the Nobel Prize.

Of course, Becquerel could not succeed in his personal endeavour without the support of colleagues and the French Academy of Sciences. This paper will not try to disclose the sociological aspects of the episode. It will only analyse the evidence relating to Becquerel's mistakes and his strategy of occultation of his own failure.

2. HENRI BECQUEREL'S EARLY WORK

Becquerel began his search for penetrating radiations emitted by luminescent bodies in January 1896. On the 24th February, he presented to the French Academy of Sciences his first positive results: he succeeded to detect a penetrating radiation (similar to X-rays) emitted by crystals of double sulphate of uranyl and potassium (Becquerel, 1896a). In this and in the next communication (Becquerel, 1896b), Henri Becquerel did not discuss the nature of the penetrating radiation. He only described that it was able to pass through black paper and thin glass plates, and to affect photographic plates. At this time, he believed that those radiations "[...] could be invisible radiations emitted by phosphorescence with a persistence infinitely larger than the persistence of luminous radiations emitted by those bodies." (Becquerel, 1896b)

Nowadays, a modern physicist will find nothing strange or unexpected in those two earlier communications. From Becquerel's third "radioactivity" paper onwards, however, he reported several phenomena that seem to us completely anomalous, in the following sense: they conflict with our physical knowledge and it is doubtful that anyone nowadays could reproduce Becquerel's observations. Becquerel and coetaneous scientists, however, found nothing strange in those phenomena. Indeed, they completely accorded with Becquerel's expectations.

In his third "radioactivity" paper (9th March 1896), Henri Becquerel began the study of the properties of the radiation emitted by the uranium phosphorescent compound he was using (Becquerel, 1896c)[2]. He was guided by his expectation that the radiation was an invisible light, and by Röntgen's investigation of X-rays. One of the known properties of X-rays and ultraviolet light was their ability of discharging an electroscope. Becquerel observed that the uranium rays were also able to discherge an electroscope. Röntgen had tried to observe reflection and refraction of X-rays, with negative results. Becquerel tried similar experiments with uranium radiation – and he apparently succeeded.

3. REFLECTION OF URANIUM RADIATION

One of Becquerel's experiments seemed to show clear evidence of regular reflection of the radiation emitted by the uranium salt. He cut a concave mirror in a small tin block. This mirror was polished and produced visual images. There was, however, a small defect of the metal, and at this point the mirror could not be polished. A thin crystal of uranium and potassium sulphate was attached to the focal plate of the mirror. Below this device, Becquerel placed a photographic plate, separed from the crystal by paper.

Becquerel reported that when the photographic plate was developed, he observed a triangular black figure, corresponding to the crystal flake. It was surrounded by a dark circle, and in this circle there was a spot corresponding to the defect of the mirror. A hypothetical reconstruction of this experiment is shown in Fig. 1.

---

[2] At this time, Becquerel did not use the phrase "uranium rays" and there was no evidence that the radiation he was studying was peculiar to uranium compounds.

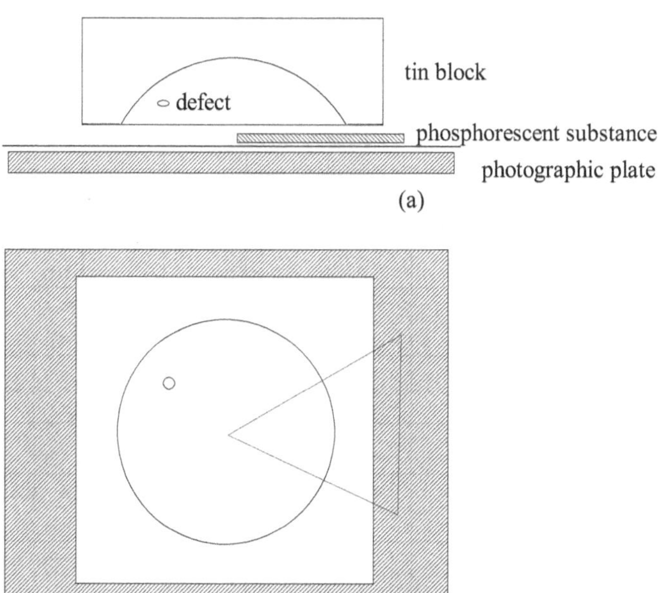

Fig. 1. A hypothetical reconstruction of Becquerel's experiment on the reflection of uranium rays.

Becquerel concluded:

> This halo, with a quite sharp border, is therefore due to radiations that, after being reflected on the mirror, were sent to the plate in nearly parallel directions. (Becquerel, 1896c, p. 561)

This seemed a conclusive evidence of regular reflection (not diffusion or scattering) of radiation. According to our present-day knowledge, uranium radiation does not suffer reflection in polished metal. Besides that, Becquerel's argument is wrong. Only if the source of radiation were very small and if it were placed at the focus of an spherical mirror, the radiation could be reflected as parallel rays. However, in the case of an extended, large source (as was the case in Becquerel's experiment), even if there were specular reflection, rays emitted from different points of the substance would have different directions after reflection and there could never arise any spot corresponding to the defect of the mirror. It is impossible to understand what happened in this experiment.

4. REFRACTION OF URANIUM RADIATION

In the same paper presented on 9th March 1896, Henri Becquerel described evidence for the existence of refraction of the penetrating radiation emitted by phosphorescent compounds. Becquerel first tried to detect refraction of uranium radiation using a prism, and stated that those experiments "gave signs of refraction, but the signs were too weak to be presented today. Moreover, it will be seen from results that will be described below, that some images clearly reveal the fact of refraction and total reflection in glass" (Becquerel, 1896c, p. 561).

The positive evidence referred to by Becquerel was obtained in the study of uranium nitrate. This substance strongly absorbs moisture from the air and its crystals must therefore be protected from the atmosphere. Henri Becquerel put the uranium nitrate in a glass tube, closed with a thin glass plate (0.2 mm thick) sealed with paraffin (Fig. 2). The crystal sample was several milimeters high.

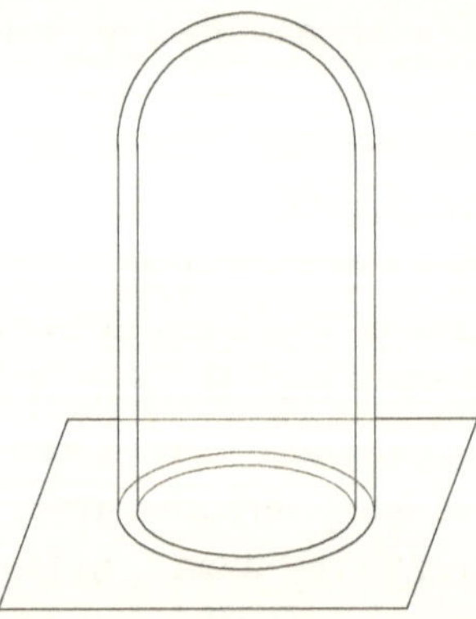

Fig. 2. A hypothetical reconstruction of the glass tube used by Becquerel in his experiment with uranium nitrate.

This device was put (the glass plate downwards) over a photographic plate wraped in black paper. After two days, the photographic plate was developed and showed a black spot corresponding to the base of the uranium nitrate crystal. This spot was surrounded by a "slightly dark" band, limited by the border of the glass tube. Becquerel concluded:

> This band is due to the action of the radiations obliquely emitted by the vertical faces of the [cylinder of powdered] crystal which is several milimetres thick; the radiations stoped by this tube were refracted and totally reflected inside it, as light rays inside a liquid vein. The action is stronger at the places that are in contact with the uranium nitrate crystal. (Becquerel, 1896c, p. 563)

Becquerel *expected* uranium radiation to be refracted and reflected, because he thought it was some kind of penetrating ultraviolet light. What he observed confirmed his expectations. However, we know that uranium radiation is not refracted or reflected by glass. This anomalous effect cannot be explained according to our present knowledge.

Besides that, in a later paper, Becquerel described that he obtained deflection of the radiation of uranium nitrate using a crown glass prism (Becquerel, 1896d, p. 693). Becquerel never published the photographic evidence of those experiments.

5. EMISSION OF PENETRATING RADIATION BY CALCIUM SULPHIDE

In his first two "radioactivity" papers, Henri Becquerel had studied the radiation emitted by double sulphate of uranyl and potassium. In his third (9th March 1896) paper of this series (Becquerel, 1896c), he described for the first time some attempts to detect penetrating radiations emitted by other substances. In a first series of experiments, he tried double sulphates of uranyl and sodium, of uranyl and potassium, of uranyl and ammonium, uranium nitrate, and zinc sulphide.

Those substances were put over thin glass plates over the same photographic plate, without exposition to any strong light. All uranium salts produced similar photographic effects through black paper. The zinc sulphide sample produced no effect. Before Becquerel's experiments, both Charles Henry (1896) and Louis Joseph Troost (1896) had observed strong effects with zinc sulphide, but in their experiments there was stimulation by sunlight or magnesium light.

In another series of experiments, Becquerel tried another set of substances: orange calcium sulphide,[3] green strontium sulphide, blende (zinc sulphide), blue calcium sulphide and greenish blue calcium sulphide. Some of those substances were altered by the air, and therefore he was led to enclose them in glass containers similar to those used for uranium nitrate. Between those tubes and the photographic plate (wrapped in black paper) there was an aluminium plate, 2 mm thick. This set of samples was exposed to ambient light (not directly to sunlight) and left over the photographic plate from 4 p. m. on the 7th March to 9:30 a. m. on the 9th March 1896 – that is, 41½ hours[4]. Only two of those substances produced observable effects on the photographic plate: the blue and greenish blue calcium sulphides. The effects produced were very strong – even stronger than those obtained with uranium compounds:

> [...] the two blue and greenish blue luminous calcium sulphides gave very energetic actions, the most intense that I have yet obtained in those experiments. The fact relative to the blue calcium sulphide accords with the observation of Mr.

---

[3] The colours described here refer to the light emitted by those substances in darkness, after excitation by light. The colour of light emitted by phosphorescent bodies usually depend on the presence of impurities.

[4] In different publications, Becquerel provides different numbers, varying from 43 to 48 hours. The exact times of the experiment are only found in the caption accompanying the image published in 1903.

Niewenglowski [that the radiation passes] through black paper. (Becquerel, 1896c, p. 563)

Besides confirming Gaston Niewenglowski's earlier observation (Niewenglowski, 1896)[5], this experiment was relevant for another reason: it provided evidence for reflection and refraction of the penetrating radiation:

> The images I have obtained with the two calcium sulphides through aluminium are worth pointing out as offering very important peculiarities. The quantity of phosphorescent powder contained in the tubes formed a column several milimetres high over the plane glass slide basis, and about one centimetre for the blue sulphide. The radiation of the lateral surface produced large black spots, exceedingly strong, in the middle of which it was possible to distinguish a clearer image of the section of the glass tube, and especially the very neat edge of the glass plate. Those edges, black inside and surrounded by an absolutely white line, show that the oblique radiations have penetrated the glass plate and have been refracted and completely reflected there at the separation surface between glass and air. Both calcium sulphide tubes presented the same appearance, in different degrees, and the radiations have even attained the nearby tube containing strontium sulphide and produced the appearance, with the same character, of part of this tube and of the slide that supported it. If the phenomena of refraction and reflection had not been evidenced before by other experiments, it would be made manifest by this single test. (Becquerel, 1896c, pp. 564-565)

Hence, this was the strongest evidence provided by Becquerel for reflection and refraction of the invisible radiation. Notice that he made no distinction here between uranium radiation and the radiation presumably emitted by calcium

---

[5] See MARTINS, Roberto de Andrade. A pool of radiations: Becquerel and Poincaré's conjecture, published in this volume.

sulphide. This is a strong evidence that, at this time, he still accepted Poincaré's conjecture as true and thought his experiments were equivalent to those of Gaston Niewenglowski, Charles Henry and others.

Most of Becquerel's work used the photographic method of detection of radiation. It is likely that Becquerel presented at the meetings of the Academy of Sciences the negatives he obtained, but they were not published in the *Comptes Rendus*. Several years later, after the appearance of the works of Marie Curie, Ernst Rutherford and others, when Becquerel's preliminary work grew in importance, he published his early photographic evidence concerning calcium sulphide (Becquerel, 1900, p. 49, fig. 1; Becquerel, 1902, plate 2, fig. 4; Becquerel, 1903a, plate II, fig. 5). There is, however, an older copy of Becquerel's photograph, sent by himself to Lord Kelvin in 1897, that deserve notice.

In the beginning of 1897, Lord Kelvin began a series of researches on uranium radiation. He was specially surprised with the electrical properties of the uranium radiation, as he told Stokes:

> Two days ago I received from Moissan a specimen of Uranium, and have seen with my own eyes its effectiveness in discharging an electrified conductor which is more like magic than anything I have ever seen or heard of in Science.[6]

Kelvin sent a copy of his first paper (Kelvin, Beattie & de Smolan, 1897) to Henri Becquerel, who replied to him on the 3rd August 1897. Becquerel enclosed with his letter two copies of the photographs of his experiment[7]:

---

[6] Letter from Lord Kelvin to Stokes, 25th February 1897. Original. manuscript kept at the Cambridge University Library, CUL Add 7656.K328. Other manuscripts consulted at this library will hitherto be identified as CUL.

[7] Letter from Henri Becquerel to Lord Kelvin, 3rd August 1897, CUL Add 7342.B52.

I send you two facsimile prints of the negatives I have obtained last year, at the very beginning of my researches.

One is the print of an aluminium medal traversed by the rays emitted by double sulphate of uranium and potassium.

The other was obtained with a powdered phosphorescent calcium sulphide that later showed itself inert. The phosphorescent powder was enclosed in a small glass tube that rested upon a glass slide sealed with paraffin, and separed from the photographic plate by an aluminium sheet 2 milimetres thick. The print shows the refraction and the total inflexion [of the radiation] at the edges of the glass slide and the paraffin.

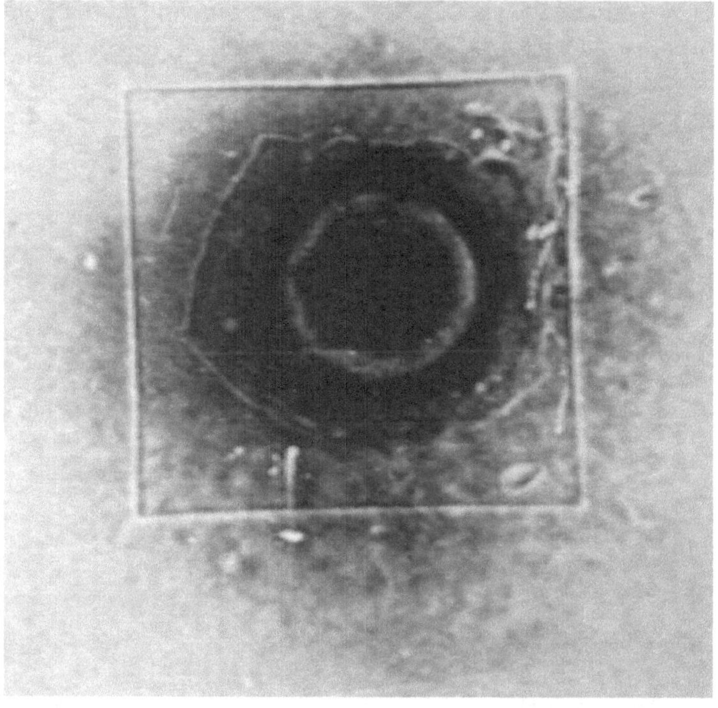

Fig. 3. Detail of the second photograph sent by Becquerel to Lord Kelvin in 1897 (CUL Add 7342.B52).

Part of the second photograph is reproduced in Fig. 3. Notice that at this time Becquerel still accepted as valid and relevant

117

his observations of the calcium sulphide samples and the evidence for the refraction of radiation. For this reason, the second photograph is the most relevant for the present discussion.

At the back of this photograph, Becquerel sketched the experimental setup and wrote:

> Facsimile of a print obtained the 7th March 1897 with the rays emitted by a preparation of phosphorescent calcium sulphide, through an aluminium sheet 2 milimetres thick.
>
> [signed] H Becquerel

Then follows the sketch of the experiment, and, at the bottom of the print: "Offered to Lord Kelvin by Mr. H. Becquerel".

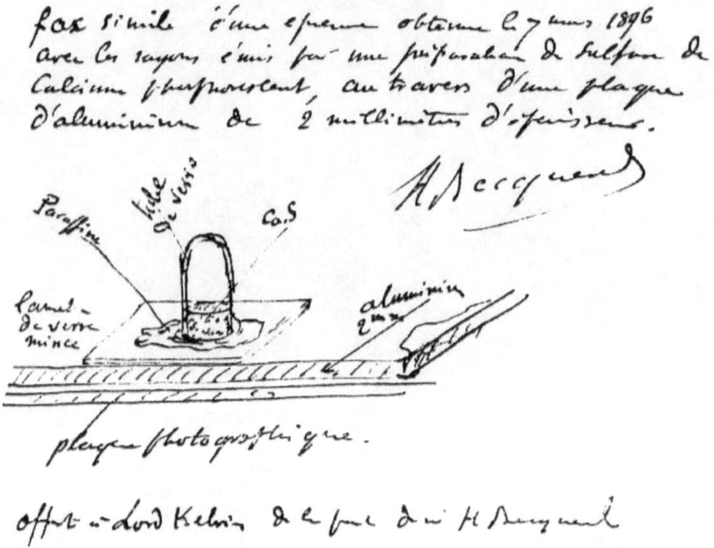

Fig. 4. Becquerel's sketch of the experiment with calcium sulphide (CUL Add 7342.B52).

Both photographs sent by Becquerel to Kelvin measured 9x12 cm. This is the exact size of Becquerel's original Lumière plates (Becquerel, 1903a, p. 42). Of course, they have been reproduced by a contact method. The experimental setup is also shown in a photograph published by Becquerel (Becquerel, 1902, plate 2, fig. 3; Becquerel, 1903a, plate II, fig. 7). The photograph shows indeed the details described by Becquerel and, at this time, was compatible with his interpretation. Becquerel seemed particularly proud of this evidence, and he provided copies of the refraction photograph to other researchers (Bouty, 1896, p. 615).

In later works, after the dismissal of refraction of uranium radiation by other researchers, Becquerel ascribed his mistake to this single calcium sulphide experiment. In his 1902 account, Becquerel describes this experiment and comments: "Those facts and some other had led me to think that the new radiation could be a transversal motion of the ether analogous to light; the absence of refraction and a large number of different experiments made me give up this hypothesis". In his 1903 book, Becquerel also attributes his mistakes to this anomalous experiment: "Unfortunately, the assimilation [of the calcium sulphide effects] with the effects produced by uranium rays and the appearance of reflection and refraction effects have led me at that time to attribute to the uranium rays properties analogous to those of light – properties that they do not have" (Becquerel, 1903b, p. 51).

However, as shown above, Becquerel had described a similar effect he observed in an experiment using uranium nitrate. The calcium sulphide experiment was not the only evidence he presented for refraction of the penetrating radiations.

The emission of penetrating radiation by calcium sulphide observed by Becquerel cannot be explained by our physical knowledge. In a later communication, Becquerel reported that the samples of calcium sulphide that had formerly given strong effects in previous experiments were now inactive. He tried to stimulate those samples by light, heat and cold, with no results (Becquerel, 1896d). A similar phenomenon had occurred with the samples of zinc sulphide used by Troost: recently prepared blende produced penetrating radiation when excited by magnesium light, but the activity gradually decreased and after some time the phosphorescent substance was unable to react to illumination (Troost, 1896).

After his samples of calcium sulphide died out, we might expect that Becquerel would perceived that the phenomenon he was studying was peculiar to uranium compounds. However, even several months after this time, he still mentioned his calcium sulphide observations as related to his uranium experiments.

6. PERSISTENCE AND STIMULATION OF EMISSION OF THE INVISIBLE RADIATIONS

There was a conflict between Becquerel's expectations and his observations concerning the persistence of the invisible radiations emitted by uranium salts. He observed that the substances he was using emitted the penetrating rays for a long time, when kept in the dark. Nowadays, we believe that this is one of the main characteristics of radioactivity: it is a spontaneous emission of radiation, that cannot be increased or decreased by common physical stimuli (light, heat, etc.). In the case of uranium, the emission decreases very slowly in time – a decrease that cannot be detected in a few years.

In his third "radioactivity" paper (Becquerel, 1896c), Henri Becquerel described the long persistence of the invisible radiations emitted by the phosphorescent crystals of uranium compounds, that he had kept in darkness for 160 hours. During this time, there was no perceivable decrease of the penetrating radiation. However, this observation did not led him to the conclusion that this was a new phenomenon:

> Perhaps this fact should be compared to the indefinite conservation of absorbed energy in some bodies, that emit it when one heats them, a fact to which I have already called the attention in a work on the phosphorecence by heat. (Becquerel, 1896c, pp. 562-563)

Henri Becquerel was still being guided by his knowledge of luminescence phenomena. The phenomenon he recalled here had been well studied by his father. When a phosphorescent substance is exposed to light and brought to a dark room, it will shine during some time, but its luminosity will decrease and after a longer or shorter time it will seem to have lost all its phosphorescence. However, there are several phosphorescent substances that can shine again after becoming dark, if they are heated, a phenomenon that had been studied by his father Edmond Becquerel:

> Once the phosphor is exposed to light and placed in the darkness, the change acquired under the influence of radiation will remain for some time, even when the sulphide does not sensibly shine any more at room temperature, and an increase of temperature can afterwards give rise to a light emission. (Edmond Becquerel, 1848)

Five years before his "radioactivity" researches, Henri Becquerel had also studied those phenomena:

> From the moment when they are submitted to the exciting action of light onwards, phosphorescent bodies kept at a constant temperature emit light that ceases being perceptible after a shorter or longer time, varying from a fraction of a second to several days, and then the body extinguishes itself. If we then rise the temperature and keept it constant again, the body becomes luminous, then extinguishes itself again; [...] so, for a given temperature, there is, on one side, a faster or slower lost of energy by light radiation, and, on the other side, an amount of energy that remains in the body in a latent state, to be emitted at a higher temperature. This latent portion of light stored in the body seems to stay therein in a permanent way, if that body is kept at a temperature equal or smaller to the regarded temperature. (Becquerel, 1891, pp. 561-562)

In the same paper, Becquerel remarked:

> There is a fact worth calling our attention, the indefinite conservation in bodies of an amount of energy that they absorb and that they emit when they are heated. By what mechanism is this energy kept without any sensible lost? Is there a particular state of matter comparable to that of magnetized bodies? Is the lost of energy continually compensated? Those are questions that we cannot answer now and that may perhaps be elucidated by future studies. (Becquerel, 1891, p. 563)

There is one sentence in the above citation that deserves carefull analysis: "Is the lost of energy continually compensated?". According to Lommel's theory, the vibrations of the particles of solid luminescent bodies are submitted to resistive forces (see Martins, 1997). There should therefore be a continuous energy lost, and if those bodies keep for a very long time (or indefinitely) their ability of emitting light when heated, this loss had to be compensated. It is very difficult to imagine any process of this kind, but the above citation shows that in 1891 Henri Becquerel considered the possibility of luminescence phenomena with continuous loss of energy and continuous compensation of this loss.

7. OTHER ANOMALOUS PROPERTIES OF BECQUEREL'S RAYS

Further experiments made by Becquerel provided new anomalous phenomena: (a) the intensity of the radiation emitted by uranium salts increased when they were stimulated by light; (b) the radiation exhibit polarization effects. Of course, according to present physical knowledge those effects could not exist, but Becquerel reported them an they strengthened the belief that the phenomenon was a kind of insivisible phosphorescence and that the emitted radiation was invisible electromagnetic radiation (similar to ultraviolet rays).

7.1 *Excitation of emission of radiation by light*

When Henri Becquerel reported the emission of radiation by his phosphorescent samples kept in darkness, he concluded that it could be due to some kind of invisible, long lived phosphorescence. In his third "radioactivity" paper, he reported that the effect was still observed when the samples were kept in darkness for 7 days. In his communication presented on the 23rd March 1896, Becquerel presented new evidence:

> If the phenomenon of emission of invisible radiations that we study is a phosphorescence phenomenon, it should be possible to exhibit its excitation by given radiations. That research becomes very difficult because of the prodigious persistence of the emission when those bodies are kept in darkness, protected from all luminous radiations and from invisible radiations of known nature. After more than 15 days, uranium salts still emit radiations almost as intense as on the first day. Placing on the same photographic plate, with black paper, a flake kept for a long time in darkness and another that had just been exposed to daylight, the impression of the silhouette of the second is a little bit stronger than the first. Magnesium light, in the same conditions, produces only an imperceptible effect. If the flakes of double sulphate of uranyl and potassium are lively illuminated by an electric arc, or by the bright sparks of the discharge of a Leyden bottle, the impressions are noticeably darker. Therefore the phenomenon seems indeed an invisible phosphorescence phenomenon, but it does not seem intimately related to the visible phosphorescence and fluorescence. (Becquerel, 1896d, p. 691)

Becquerel was not the only one who reported this effect. Silvanus Thompson also stated that stimulation by light increased the emission of penetrating rays by uranium nitrate (Thompson, 1896a, p. 713). In the case of metallic uranium, Thompson stated that "the hyperphosphorescence of uranium in the metallic state is about equal in darkness and when exposed to light" (*ibid.*).

Becquerel's study of increased emission required the visual comparison between two black spots produced by different phosphorescent flakes upon photographic plates. It is very difficult to arrive at any definite conclusion if the spots are similar, and he could be misled by his theoretical expectation. However, there was an independent, objective method that could be tried: the measurement of the effect of the radiation upon the discharge of an electroscope.

Henri Becquerel also used this second method:

> The electroscope allowed me also to display the weak difference between the emission of a flake of uranium salt kept in darkness for eleven days, and the emission of the same flake vigorously iluminated by magnesium. In the first case, the speed of fall of the [electroscope] leaves was 20.69 [seconds of arc per second] and after luminous excitation it became 23.08. (Becquerel, 1896d, p. 691)

There was a second series of measurements. On the 28th March 1896, he measured the speed of discharge of an electroscope due to the action of a flake of double sulphate of uranyl and potassium. At 1:45 p.m., shortly after the uranium salt had been exposed to light, the speed was 38.18" /s; at 6:20 p.m., the speed was 33.60" /s. On the next day, at 5:40 p.m., the speed was 33.00"/s. He concluded:

> The numbers cited above show that, a short time after being exposed to light, the action of the flake of the uranium salt was a little bit stronger. In five hours there happened a slight weakening, and afterwards the action remained sensibly constant up to the next day. (Becquerel, 1896e, p. 765)

The two series of measurements are in qualitative agreement. In both cases, a decrease of the intensity of radiation larger than 10% was observed when the uranium salt was kept in darkness. There seemed to exist strong evidence for accepting the increase of radiation intensity under stimulation by light and to interpret the phenomenon as a kind of phosphorescence. In later papers, Becquerel still held the same opinion (Becquerel, 1896f).

7.2 *Polarization of Becquerel's rays*

An important property of transversal waves is the possibility of polarizing them. After the discovery of X-rays, several researchers have tried to polarize them by reflection and using tourmaline crystals. If X-rays could be polarized, this would be a great step in the search for their nature. Although some papers reported positive results (Galitzine & Karnojitsky, 1896), most observers had failed to detect polarization of X-rays (Thomson, 1896).

Henri Becquerel believed that the radiation he was studying was similar to light. He had already "proved" that it could be refracted and reflected. It was natural to check whether it could be polarized. On the 30th March 1896, he reported positive evidence for the polarization of uranium radiation (Becquerel, 1896e).

A photographic plate was wrapped in black paper. Over the paper Becquerel placed two pieces of a thin tourmaline sheet (0.50 mm), oriented in perpendicular directions. Over them he put a single tourmaline layer (0.88 mm thick), with its axis parallel to that of oe of the small tourmalines and perpendicular to the other. In this condition, light passed through the parallel tourmalines and is stopped by the crossed tourmalines. A flake of double sulphate of uranyl and potassium was placed over this device.

> After 60 hours of exposition, the photographic plate was developed; it clearly showed the silhouette of the tourmalines, and the action through the parallel tourmalines was considerably stronger than through the crosses tourmalines. [...]
> This experiment therefore shows at the same time, for the invisible rays emitted by uranium salts, the double refraction, the polarization of both rays and their different absorption through the tourmaline. (Becquerel, 1896e, p. 763)

Becquerel repeated the same experiment using X-rays instead of uranium radiation, and observed no polarization effect.

In this case, as in some others, Becquerel had very scarce experimental evidence: his polarization experiment was *a single test* (perhaps repeated once, several days later). One of the most crucial pieces of evidence for the interpretation of the nature of uranium rays was the difference between two diffuse dark spots in a photographic plate.

In his 1903 book, Henri Becquerel published for the first time his photographic evidence for polarization of uranium rays (Becquerel, 1903b, plate II, fig. 6). It is very difficult to recognize the effect described by Becquerel (Fig. 5).

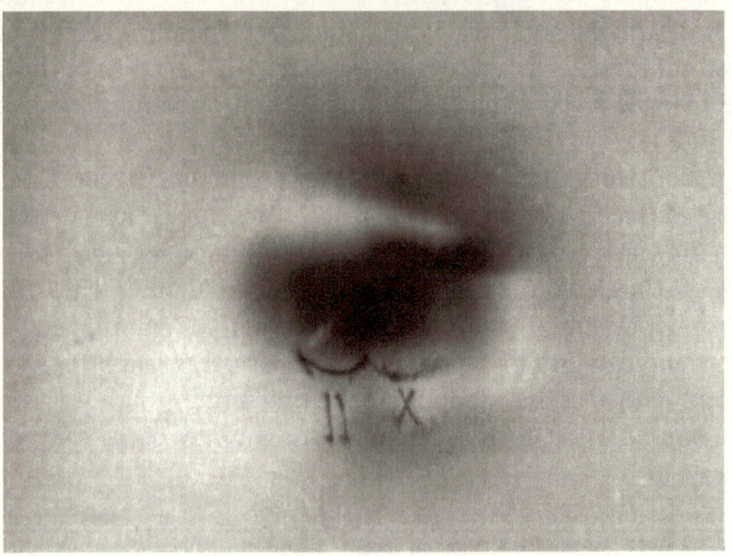

Fig. 5. Photograph of Becquerel's experiment to show the polarization of uranium rays. The region identified by II corresponds to parallel tourmaline plates, and the regions identified by X corresponds to crossed plates..

7.3 *Conclusions drawn from Becquerel's experiments*

At this time, Becquerel's experiments seemed to clearly prove that the radiation emitted by uranium compounds was a kind of electromagnetic transversal wave. Silvanus Thompson discussed the nature of the uranium radiations, and remarked:

> The extraordinary property exhibited by the uranium compounds of emitting a persistent invisible radiation that will pass through aluminium and produce photographic action would suggest that these rays are identical with Röntgen's, were it not that Becquerel's success in reflecting, refracting, and polarizing them proves that they are more akin to ultraviolet light. (Thompson, 1896b, p. 107)

8. RECTIFICATION OF BECQUEREL'S MISTAKES

Before 1898, Becquerel's work was not submitted to systematic duplication or criticism. It was simply reviewed and accepted as a contribution that did not strongly contrast with other known phenomena and therefore called for no deeper thoughts. The presentation speech of Becquerel's Nobel Prize completely misrepresents the historical situation, when refers to the reception of Becquerel's work, *before the work of Pierre and Marie Curie*, in the following words:

> It goes without saying that a discovery such as this was bound to excite the liveliest interest in the scientific world and give birth to a whole host of new investigations with the aim of making a thorough study of the nature of the Becquerel rays and determining their origin. (Törnebladh, 1967, p. 48)

Before 1898, only two aspects of Becquerel's work had been criticized: the polarization of uranium rays and the excitation of this radiation by light. Gustave le Bon, whose work on "black light" had been strongly criticized by Henri Becquerel, was the first to deny the polarization of uranium radiation, in May 1897:

> I suppose that the qualification "pretended" applied to black light means simply that my experiments do not always succeed. They do indeed not invariably succeed mainly because of the difficulty in preparing plates that are sensible

to those new radiations. But if the qualification "pretended" should be applied to all experiments that do not always succeed, couldn't we apply it also to uranium rays? Would Mr. Becquerel like to state, for instance, that his experiments on the polarization of uranium rays – the polarization that if demonstrated would have a fundamental importance concerning the nature of those rays – can be repeated at will? Will he achieve repeating them safely? I have some reason to doubt it. (Le Bon, 1897, p. 691)

Notice that in his comment Le Bon did not clearly state that uranium rays cannot be polarized. He presented a *doubt*, not a clear *denial*. He also didn't state that he had repeated Becquerel's experiments. In January 1899, however, he clearly stated that he repeated the experiments and that uranium radiation cannot be polarized:

Metallic radiations, including uranium radiations, have never shown any trace of polarization, by any of several methods I have used, even after an exposure of three months. It is certain, therefore, that those radiations [...] that cannot be a kind of light, are a new form of energy, as I have formerly said. (Le Bon, 1899, pp. 108-109)

Of course, the non-observation of polarization does not *prove* that something is not an electromagnetic radiation: recall that no polarization effect of X-rays had been observed. However, the denial of Becquerel's supposed prove of polarization of uranium rays destroyed one of the strongest arguments for their interpretation as high frequency ultraviolet rays.

In his first paper on the radiation emitted by thorium, Schmidt believed that he had found evidence for refraction, but found no sign of polarization by tourmalines (Schmidt, 1898).

In the beginning of 1899, Ernest Rutherford reproduced Becquerel's polarization experiment and could not perceive "the slightest difference in the intensity" of radiation passing through parallel or crossed tourmalines (Rutherford, 1899, p. 112). In the same paper, Rutherford described experiments to test refraction of uranium rays. He used prims of glass, aluminium and paraffin. The prisms were crossed by uranium radiation emerging from a slit cut in a thick lead plate. He observed no deflection of the radiation.

Johann Philipp Elster and Hans Friedrich Geitel found, in 1897, that the emission of uranium radiation could not be increased by excitation by sunlight (Elster & Geitel, 1897). The intensity was found to be constant (not "slightly decreasing", as Henri Becquerel described it) over several months. In this paper, Elster and Geitel stressed that the radiation from uranium can be distinguished from the effects produced by other substances (aluminium, zinc, phosphorescent paind and fluorspar) because these do not impart electrical conductivity to the air. Notwithstanding the title of their paper, they concluded that the name "hyperphosphorescence" cannot be applied to the observed phenomenon. This seems the first time that the concept of an invisible phosphorescence of uranium was criticized.

In April 1898, thorium was discovered to emit radiations similar to those of uranium (Badash, 1966). This led to an increased interest in the phenomenon. In this same year, polonium and radium were also found. In 1898, Marie Curie also rejected the name "hyperphosphorescence" and proposed the name "radioactivity":

> Uranium rays have frequently been called *Becquerel rays*. This name can be generalized and applied not only to uranium rays but also to the rays of thorium and to all similar radiations.
>
> I will call *radioactives* the substances that emit Becquerel rays. The name *hyperphosphorescence* that had been proposed for the phenomenon seems to me to convey a wrong idea about its nature. (Curie, 1899, p. 50)

The Curies rejected the old "invisible phosphorescence" concept, but proposed an explanation of radioactivity related to invisible fluorescence. Indeed, for several years they claimed that there existed an unknown, invisible, very penetrating cosmic radiation (similar to extremely hard X-rays), that could be transformed by radioactive bodies into less penetrating, detectable rays, by a process similar to fluorescence.[8]

The Curies concentrated their attention on the substances emitting radiation and not on the radiations. In 1899, Rutherford identified two kinds of radiation (called by him α and β) using as criterion the absorption of radiation by thin aluminium foils (Rutherford, 1899). A few months later, Fritz Giesel (1899) showed that β radiation could be deflected by a magnet and therefore could not be an electromagnetic radiation (see Malley, 1971). After a few years, a completely new view emerged: radioactive bodies emitted three kinds of radiation, two of them (α and β) deviable by magnetic fields (and, therefore, carrying electrical charges), and the third (γ), non deviable, similar to X-rays. The nature of the radiation emitted by uranium and other radioactive bodies was completely different from what Becquerel had believed and "proved" by his experiments.

The central core of our present theory of radioactivity was built by Rutherford and Soddy in 1902-3. They presented strong evidence for the gradual transformation of radioactive elements, the existence of radioactive series and spontaneous release of internal energy (Rutherford & Soddy, 1902a; 1902b; 1903; cf. Malley, 1979; TRENN, 1975).

9. BECQUEREL'S STRATEGY

In 1899, Henri Becquerel acknowledged for the first time some of his early mistakes, but tried to convey the impression that he had corrected them himself (Becquerel, 1899). From this time onwards, he devoted much of his energy to establish himself as the successful discoverer of radioactivity.

---

[8] See MARTINS, Roberto de Andrade. The guiding hypothesis of the Curies' radioactivity research: secondary X-rays and the Sagnac connection, in this volume.

It is remarkable that, at one point of his 1903 book, which presented the state of the art of radioactivity up to that time, Becquerel stated that his only aim was to describe his own researches: "To describe the beautiful work of Mr. and Mrs. Curie is outside the scope of this memoir, that in principle contains only my personal researches" (Becquerel, 1903b, p. 105). Maybe this meant that the researches of other people, described in his book, were accessory to his own work.

Henri Becquerel used a systematic strategy: he turned his old mistakes into as so many successes; he described as his own the discoveries of others; he distorted the whole history of radioactivity and tried to show that he was the central protagonist. Let us show some instances of this strategy.

9.1 *Spontaneity of radiation*

Before 1898, Becquerel had never described the emission of uranium radiation as "spontaneous". Afterwards, when this was seen to be one of the fundamental aspects of radioactivity, Becquerel reinterpreted his work:

> Among the properties that I have pointed out at the beginning of my researches as characteristic of this radiation that was unknown, there are three fundamental ones that have been afterwards verified by all observers; they are: the spontaneity of radiation, its constancy and the property of imparting electrical conductivity to gases. (Becquerel, 1899, p. 771)

After describing his first "radioactivity" paper, Becquerel stated:

> Under those conditions, the phenomenon could be attributed to a transformation of solar energy, of the same kind as phosphorescence, but I soon recognized that emission was independent of any excitation of known nature – luminous, electric or thermic.

> We were therefore in face of a spontaneous phenomenon of a new kind. Here I show you the first print which revealed the spontaneity of the radiation emitted by the uranium salt. (Becquerel, 1903c, p. 2)

and at this point, Becquerel mentioned the first photograph taken in darkness, described in his second "radioactivity" paper. At other places, Becquerel explicitly stated that he recognized at this time the spontaneity of uranium radiation:

> This observation establishes the fundamental new fact of an emission of penetrating rays without apparent exciting cause. (Becquerel, 1903a, p. 13)

> [...] some days later, from 27th February to 1st March, I recognized that the emission was produced spontaneously, even when the uranium salt was kept protected from luminous excitation [...]. On the 2nd March 1896, I reported to the Academy of Sciences the conditions under which I have been led to observe the spontaneity of the radiation, the new fact from which follow all later studies. (Becquerel, 1900, p. 48)

9.2 *Constancy (in time) of emission*

Up to 1898, Becquerel described that the emission of radiation by uranium salts decreased with time, after stimulation by light. Afterwards, the story was changed.

According to Becquerel, after noticing that the uranium salt emitted radiation in darkness, he already supposed that the intensity was constant:

> As the uranium salts used had been prepared a long time ago, it was to be supposed that the intensity of the phenomenon was independent of time, and hence that emission should appear constant. All later experiments showed that the activity of uranium presented no appreciable decrease with time.
>
> [...] The photographic method was primarily a qualitative one while the electrical method gave numerical data, and the

early measurements revealed the constancy of the radiation with time. (Becquerel, 1903c, p. 2)

In 1899, Becquerel still accepted that the intenstity of uranium radiation exhibited a decrease with time:

> It seems that there is a slight decrease of intensity during the first months and afterwards the intensity seems unchanged. (Becquerel, 1899, p. 772)

In this same paper, Becquerel stated that uranium radiation cannot be stimulated by physical influences, but did not acknowledge that Schmidt corrected him:

> [...] it was impossible to produce any noticeable change of the intensity of this emission by physical influences. (Becquerel, 1899, p. 777)

At other places, Becquerel stated that his early experiments showed that the intensity was constant:

> From the begining of those studies I have checked whether onc could observe a progressive weakening of the radiated energy by subtracting those bodies to all known external excitation. A first series of experiments, pursued during two months, has initially showed that this energy did not decrease in an appreciable way. (Becquerel, 1900, p. 52; cf. Becquerel, 1903a, pp. 14-15)

At some places Becquerel referred to experiments that had shown an increase of the radiation when uranium salts were excited by light, but he did not state that himself had reported those effects:

> None of the attempts to exhibit an excitation by ultraviolet, infrared or light rays produced a [positive] result; the same was the case when uranium salts were excited by X-rays. However, in several experiments, after exposing the double

sulphate of uranyl and potassium flakes to the action of sparks and electrical arc, a slight temporary increase of emission was observed, but this very weak effect seems another phenomenon superposed upon the constant and continuous emission by uranium. (Becquerel, 1900, p. 53)

Finally, in his 1903 book Becquerel stated that, using the electroscopic method, he had been able, as early as 14th March 1896, to prove that the intensity of radiation was not increased when the uranium salt was excited by magnesium light:

> In some cases, the photographic impression produced by samples of a salt exposed to light or strongly iluminated by electric sparks seemed stronger thatn the impression produced by the same bodies carefully kept away from any excitation. [...] But it seems that hose facts are accidental, because electrical measurements and experiments made in order to analyse the active rays have not allowed us to detect any action of this kind. This is, for instance, one of the earlier measurements made to detect this effect. (Becquerel, 1903a, p. 21)

Below this paragraph, Becquerel presented this table:

Double sulphate of uranyl and potassium in the electroscope, 1 cm. below the gold leaves (14 March 1896)

URANIUM SALT PROTECTED FROM EXCITATION			URANIUM SALT EXCITED BY MAGNESIUM LIGHT		
t	α	$d\alpha/dt$	t	α	$d\alpha/dt$
4h38m	15°.8		5h12m	19°.6	
5h07m	5°.8	0.34	5h39m	10°.0	0.35

The table seems, indeed, to exhibit a constancy of rate of discharge of the electroscope. Why, then, Becquerel did not publish this result in the early 1896, establishing the lack of excitation of uranium radiation by light? Well, it seems that he *did* refer to this experiment, on his communication to the

Academy of Sciences of 23rd March 1896 – but the data and conclusion were different. In that paper, Becquerel stated:

> The electroscope also allowed me to exhibit the slight difference between the emission of an uranium salt flake kept for eleven days in darkness, and the emission of the same flake strongly illuminated by magnesium. In the first case, the speed of fall of the [gold] leaves was 20.69, and after luminous excitation it became 23.08. (Becquerel, 1896d, p. 690)

If we transform those speeds from seconds of ar per second of time, to degrees per minute, we obtain respectively the values 0.34 and 0.38. The first value agrees with the above table (before excitation), but the second value is different.

It could have happened that Becquerel made several series of measurements, and that the series published in 1903 is not the same referred to in his 1896 communication. Notice, however, that the speed without excitation, computed from the data of the table, is not exactly 0.54, but $10/29$ degrees per minute, equivalent to 20.69 seconds of arc per second – exactly the value published in the 1896 paper. This coincidence suggests that the communication of 23rd March was indeed reporting the same experiment described in 1903, but either in 1896 or in 1903 Becquerel made-up one of the measured speeds.

9.3 *Reflection and refraction of uranium radiation*

In 1899, Becquerel acknowledged that other researchers had shown that uranium radiation cannot be polarized, reflected or refracted, but his description implies that he had also, independently, arrived to that conclusion:

> Of the other properties that I have mentioned, polarization, reflection and refraction have not been verified by several observers that have repeated those experiments. The observations that I have made for three years have also

disconfirmed my early conclusions and have shown that the phenomena were more complex. (Becquerel, 1899, p. 772)

In some of his accounts, Becquerel provided no hint about someone else having corrected his work:

> [...] I had been led to attribute to uranium radiation the properties of light, while all later experiments have demonstrated that this radiation cannot be reflected and refracted like light rays. (Becquerel, 1903c, p. 3; cf. Becquerel, 1902, p. 86)

Becquerel referred to the reflection experiments with metal mirrors:

> However, those experiments, and other that I do not cite here, do not allow us to conclude that there is regular reflection. I have repeated many times, with variations, one of my earlier experiments, in which a fragment of active substance was placed below a small concave tin mirror that produced nice optical images and that was adjusted to produce the image of the substance on the [photographic] plate; in this way I have obtained the print of no image and, in most cases, the mirror surface seemed the source of a new radiation, producing a stronger impression of the borders of the mirror than of the central regions that were farther from the plate. (Becquerel, 1899, p. 773; cf. Becquerel, 1903a, p. 26)

After this description, which is completely at variance with his early publications, Becquerel referred to Schmidt's work:

> The experiments of Mr. Schmidt with thorium have also led him to admit a phenomenon of diffuse reflection. (Becquerel, 1899, p. 773)

Hence, according to Becquerel, Schmidt dit not correct his mistake – he just confirmed his findings.

At other places, Becquerel described the concave mirror experiments without acknowledging the authorship of the mistakes:

> The effect, initially attributed to reflection, is due to the emission of secondary rays produced at the mirror by uranium radiation. Those rays produce diffuse prints; images like those of light rays are not obtained. (Becquerel, 1900, p. 54)

The same omission of the author of the mistake is used at several other places:

> It was observed that a flake covered by a steel mirror produces a stronger impression under the mirror than another non-covered flake. This phenomenon, initially attributed to reflection, is a secondary phenomenon to which we will return at another chapter. (Becquerel, 1903a, p. 16)

While describing his early acceptance of refraction of the uranium radiation, Becquerel usually only described his experiments with calcium sulphide and did not recall his similar experiments with uranium nitrate.

At some places, Becquerel acknowledged that Rutherford was the first to find out that uranium radiation is not refracted by glass and aluminium:

> Later experiments made by Mr. Rutherford and that I could repeat have shown that the uranium radiation was not deflected by glass, paraffin wax or aluminium prisms [...] (Becquerel, 1900, p. 54)

However, sometimes Becquerel attributed to himself the discovery that uranium radiation suffers no refraction:

> The explanation given above could only be accepted if it was verified that it is possible to deflect the studied radiation with a prism of a transparent substance. Now, experiment

shows that the radiation passes without sensible deviation through glass and aluminium prims.

These are some of the arrangements that allowed me to find out this fact: [...] (Becquerel, 1899, p. 775)

Becquerel did not state here that his prism experiments had apparently exhibited refraction, and that it was Rutherford who refuted his conclusions. He only stated that:

This experiment is analogous to one of Mr. Rutherford's experiments, that gave him the same negatif result. (Becquerel, 1899, p. 775; cf. Becquerel, 1903a, pp. 110-113)

In his 1903 book, Becquerel also implied that Rutherford had just repeated his experiments:

In 1899, nearly three years after my first publications, Mr. Rutherford methodically retook, one by one, most of the acpects that I had pointed out, and measured them. (Becquerel, 1903a, p. 93)

Mr. Rutherford, after applying the photographic method to verify the absence of refraction and polarization of uranium rays, used only the electrical method. (Becquerel, 1903a, pp. 94-95)

Of course, Becquerel should have written: "after applying the photographic method to refute the refraction and polarization of uranium rays" ...

9.4 *Polarization*

At several places, Becquerel suggested to his readers that he had disproved the polarization of uranium rays before other researchers:

One first photographic print, that I have shown to the Academy in March 1896 had presented a difference of absorption through tourmaline plates, according as they were

> crossed or parallel. A second trial obtained a few weeks later had given a similar result, but all other later experiments, whether with uranium or with radium, have been negative. Other observers have also arrived to this same result. (Becquerel, 1899, p. 772)

> In two successive tests, the impression through crossed tourmalines was smaller than through parallel tourmalines, but later experiments have not given the same results, whether with the same tourmalines or with other systems. The same negative conclusions were observed by Mr. Rutherford and Mr. Le Bon. (Becquerel, 1900, p. 54)

At other places, Becquerel omitted the works of other researchers:

> A second trial made with the same tourmalines gave a result similar to the first [...], but the effect is doubtless due to an accident, because all tests made afterwards, either with the same tourmalines or with a large number of different ones, gave negative results and both transmitted beams were equally strong. (Becquerel, 1903a, p. 25)

9.5 *Becquerel's criticism of Rutherford's work*

In the first chapter of his book, where the fundamental properties of uranium radiation are presented, Henri Becquerel describes all evidence as if he alone had arrived to all the correct conclusions: the radiation is spontaneous, is not increased by excitation by physical agents, the rays cannot be reflected, refractes or polarized, etc. No credit is given to Rutherford, Schmidt, Le Bon and others. He also conveys the view that he was the main protagonist of radioactivity research after the discovery of emision of radiation by thorium. Still worst: Henri Becquerel criticizes and detracts the works of several researchers – particularly that of Rutherford[9]. For instance:

---

[9] It would be impossible to list all evidence of systematic depreciation of other researchers in Becquerel's book. As an example, see his

> Mr. Rutherford [...] concluded, as I had previously recognized, that the radiation [of uranium] is heterogeneous. Mr. Rutherford interprets his experiments considering two parts in the radiation: one, α, stronger, but easily absorbable; the other, β, much weaker, but very penetrating; he also supposes that β radiation is almost homogeneous.
>
> We are going to see below that the phenomenon is not that simple [...]. (Becquerel, 1903a, p. 95)

Of course, this remark was unjust. Becquerel had previously ascertained that uranium radiation was not homogenous *in the same sense that the X-rays emitted by a Crookes tube are not homogeneous* (Becquerel, 1896e, p. 765). Rutherford's conclusion about the existence of two different kinds of radiations (α and β) emitted by radium was a completely different thing. Besides that, the treacherous comment that "the phenomenon is not that simple" is clearly intended to discredit Rutherford's work.

At many places, Becquerel stated that Rutherford's experiments were a later reproduction of his own work: "I should however recall that this work [Rutherford's] was made two years after the measurements I had published on the same subject" (Becquerel, 1903a, p. 100). At other places, he also called the attention of the reader to limitations of Rutherford's 1899 research: "In those experiments and the conclusions drawn therefrom, the heterogeneity of the emitted radiation beam has not been taken into account" (Becquerel, 1903a, p. 97).

While describing Rutherford's discovery of the magnetic deviation of rays, Becquerel was also critical:

> Mr. Rutherford employed with a high competency, as we see, a relatively rough electrical method to exhibit a highly delicate phenomenon.

---

description of the works of Giesel, Meyer and Schweidler on the magnetic deviation of radiation (Becquerel, 1903a, pp. 126-34).

> However, the experimental disposition rises a serious objection. [...]
>
> The conclusions of Mr. Rutherford could therefore be put in doubt, if they were not confirmed by other experiments. (Becquerel, 1903a, pp. 187-188)

Rutherford read Becquerel's book and was not very happy with what he found there. He wrote letters to several of his friends, asking their oppinion about Becquerel's book. Soddy wrote to him:

> I have just got Becquerel's book, but beyond the impression he gives that he, Becquerel, is a very ignorant person, have not yet formed any opinion of it.I don't think it is of any account.[10]

Oliver Lodge also replied to Rutherford:

> I have indeed not read Becquerel's book. I have hardly glanced at it.
>
> I know that Frenchmen have a great tendency to write so as practically to claim everything for France. It seems to be the fashion there. [...]
>
> I think that in this country your reputation is safe.[11]

In the long run, Rutherford's work became widely recognized, but in 1903, after the publication of Becquerel's book and his nomination for the physics Nobel Prize, Rutherford might have felt that his researches could be misinterpreted or forgotten. It was probably in order to present his own version of the story – the one generally accepted nowadays – that he immediatly began to write his famous book *Radioactivity* (Rutherford, 1904).

---

[10] Letter from Frederick Soddy to Ernest Rutherford, 12th December 1903. CUL Add 7653.S117.

[11] Letter from Oliver Lodge to Ernest Rutherford, 11th December 1903. CUL Add 7653.L109.

This episode shows that scientists should never be trusted as faithful historians of their own work and that even a famous researcher may have been the actor of a comedy of errors.

10. HENRI BECQUEREL AND N-RAYS

In 1903, René Prosper Blondot (1849-1930) "discovered" a new kind of radiation while studying X-rays. He called them "N-rays", as a homage to his home city, Nancy (Blondlot, 1903a, 1903b). The story of this episode has been described by several authors (Nye, 1980; Lagemann, 1977; Klotz, 1980; Rosmorduc, 1972; Martins, 2007).

Blondot claimed the observation of a penetrating radiation that could not be detected by some other physicists. The method of detection usually involved the visual observation of delicate changes of intensity of electric sparks and phosphorescent screens. Such method is strongly affected by subjective factors, such as suggestion and expectations. In 1904, the lack of reproductibility of Blondot's results led to strong attacks by the scientific community and by 1905 the vast majority of researchers believed that the N-rays did not exist.

Most of the work on N-rays was developed at Nancy. However, some researchers from other places visited Blondot's laboratory and afterwards carried out investigations on those radiations. One of them was Jean Becquerel (Nye, 1980, pp. 141, 152), the only son of Henri Becquerel – the fourth generation of the Becquerel scientific dynasty.

Shortly after the "death" of N-rays, Jean Becquerel's work was strongly criticized:

> Mr. Jean Becquerel was compeled to go to Nancy to succeed, but afterwards he threw himself with passion in a series of researchers that rendered N-rays as suspect for physicists as those of Mr. Charpentier for physiologists. [...] And this rendered still less admissible the results of Mr. Charpentier, and those of Mr. Jean Becquerel, who successfully chloroformed metals, and noticed their

anaesthesia by the diminution of emission. (Piéron, 1907, pp. 148-149)

Some recent authors have supposed that Jean Becquerel was a "reputed scholar" (Rosmorduc, 1972, p. 20) and that he had his own research laboratory at this time (Klotz, 1980, p. 131). As a matter of fact, in 1903 Jean Becquerel was a 25 years old engineer who had just become his father's assistant at the Paris Museum of Natural History[12]. He had never published a single article before his papers on N-rays. Jean Becquerel was not an experienced scientist working on his own laboratory – he was certainly working at his father's laboratory, under Henri Becquerel's supervision. Jean Becquerel's communications on N-rays to the Paris Academy of Sciences were not read by himself – they were presented by his father. Although the explicit authorship of the N-rays papers did not include the name of Henri Becquerel, he completely assumed their content.[13] Indeed, in December 1904, when most of the scientific community had strong doubts concerning the reality of N-rays, Henri Becquerel presented this opinion on the subject:

> Mr. Becquerel declares that his opinion on the question is well known, from the notes of Mr. Jean Becquerel that he communicated to the Institute, and that at this moment he, as well as his son, has nothing to change therein. (Becquerel, 1904, p. 718)

There is further evidence of Henri Becquerel's support of N-rays. In 1904, he was a member of the Lecomte Prize commitee of the Academy of Sciences. He prepared a positive report on Blondot's work, including N-rays, and defended that the Prize

---

[12] The four physicists of the Becquerel family have successively occupied the same physics chair at the Museum of Natural History.
[13] It is likely that Henri Becquerel preferred to present those papers in the name of Jean Becquerel to give his young son the opportunity of recognition by the scientific community.

should be given to Blondot (Piéron, 1907, p. 153)[14]. Besides that, in his course of 1904-1905, Henri Becquerel lectured on X-rays, uranium rays and N-rays (Nye, 1980, p. 1545; Piéron, 1907, p. 161).

Blondot was smart enough to use Becquerel's support to strenghten his claims:

> In Paris, Mr. Jean Becquerel has published in the *Comptes Rendus* many notes on N and N_1 rays that were presented to the Academy of Sciences by his father, Mr. H. Becquerel, the eminent physicist. Is it possible that Mr. Jean Becquerel, with the concordance of his father, risked to endanger one of the most illustrious names of science, publishing observations that would leave the least doubt? (Blondlot, 1904, p. 621)

Why did Henri Becquerel accept as decisive proof the elusive visual observation of flickering sparks and phosphorescent screens? And why did he get so deeply involved with N-rays?

The first question is easly to answer: this was just a repetition of Henri Becquerel's mistakes in his early work on uranium radiation, when he accepted as decisive proof of delicate effects an irregular spot on a photographic plate. George Stradling commented on the effect of expectancy on N-rays research:

> A perusal of the literature on the N rays leads to the thought that the investigators often had the satisfaction of finding what they expected to find. (Stradling, 1907, p. 186)

Once Henri Becquerel was convinced that a phenomenon should exist, any dim effect was sufficient proof for him. In the case of uranium radiation, we have already discussed the

---

[14] Henri Becquerel's initial report was replaced by another one written by Henri Poincaré, who circumvented the delicate situation by proposing that the prize should be given to Blondot because of all his scientific contributions.

theoretical expectations that guided his work. What about the N-rays? This leads us to the second question.

Let us examine Blondot's first paper "On a new kind of light" (Blondlot, 1903a), presented at the Paris Academy of Sciences meeting on 23rd March 1903. Blondot studied the radiation emitted by an X-ray focus tube, using as detector a small spark jumping between the points of two wires connected to a high voltage source (usually the same induction coil used to produce the X-rays). With this detector, Blondot claimed to detect polarization, reflection, refraction and diffusion of the radiation and therefore he concluded that they could not be X-rays – they should be a new kind of rays.

> From all that precedes it results that the rays that I have thus studied are not those of Röntgen, since the latter do not suffer refraction or reflection. Indeed, the small spark reveals a new kind of radiations emitted by the focus tube: those radiations pass through aluminium, black paper, wood, etc.; they are plane-polarized when they are emitted, they are susceptible to circular and elliptical polarization, they are refracted, reflected, diffused, but produce no fluorescence, nor photographic action. (Blondlot, 1903a, p. 737)

Let us recall that Becquerel had claimed the detection of reflection, refraction and polarization of uranium radiations. After the retification of those claims by other researchers, Becquerel wrote:

> One should therefore conclude that the most active part of the uranium radiation in those experiments suffers no reflection, no refraction and cannot be polarized as light. [...] But it has not been proved that this radiation is not accompanied by radiations identical with those of light.
>
> Future experiments will probably bring us the explanation of the contradictory phenomena that have just been pointed out. (Becquerel, 1900, pp. 54-55)

In his first communication on N-rays, Blondot remarked that his new radiation could provide the explanation of Becquerel's anomalous results:

> It is interesting to approximate the foregoing to the opinion issued by Mr. Henri Becquerel that, in some of his experiments, "appearances identical with those produced by refraction and total reflection of light could have been produced by luminous rays that had crossed aluminium". (Blondlot, 1903a, p. 738) [15]

Of course, Henri Becquerel did not assume that common (visible) light can pass through aluminium, but conjectured that radiations similar to light (polarizable, reflectable and refractable) could have produced the anomalous effects he described in his early experiments. Now, in 1903, Blondot was offering him a radiation suitable for the explanation of those anomalies.

There is one particular paper by Jean Becquerel that reminds us of Henri Becquerel's old ideas on the spectrum of uranium compounds. Blondot had measured the wavelengths of N-rays; Jean Becquerel searched – and found – a periodicity in this spectrum, as Henri Becquerel had done for uranium compounds.

> It seems possible to look for the origin of N and N_1 rays in the molecular motions that are produced in all bodies in state of deformation or molecular transformation. [...]
> Those considerations have led me to examine whether the wavelenghts measured by Mr. Blondot in the beam emitted by a Nerst lamp would present simple relations between them, as the wavelenghts of the motions produced by vibrating bodies. (Jean Becquerel, 1904, p. 1332)

The remarkable similarity between this work and Henri Becquerel's work on uranium spectral bands (Becquerel, 1885;

---

[15] Blondot referred to Becquerel (1901).

see Martins, 1997) suggests that the same theoretical considerations were behind both studies.

All this suggests that Henri Becquerel's first stimulous for the study of N-rays was the expectation that those radiations could provide a suitable explanation for his anomalous results. As in the case of uranium radiation, his preconception intereferred with his observations and led him to describe non-existent phenomena.

Jean Becquerel published 11 communications between May and August 1904. After the "death" of N-rays, Jean Becquerel did not acknowledge his mistakes (Piéron, 1907, p. 154). During one year he published nothing and then turned, under the direction of his father, to "normal" researches.

11. ANALYSIS OF BECQUEREL'S MISTAKES

Becquerel was the author of a variety of experimental error. Most of them cannot be classified among the usual kinds of laboratory mistakes[16].

In order to discuss Becquerel's mistakes, it will be necessary to depend on the use of current scientific knowledge. There is nowadays some resistance against the use of anachronic knowledge within history of science, but there are also good arguments for their use, in this case: it allows reconstruction of the object of investigation (see Pickstone, 1995). It could happen, in principle, that Becquerel was never wrong, that all his experiments were correctly done and successful, but due to nationalistic preconceptions scientists of other countries (England, Germany, etc.) criticized him and were able to convince the scientific community that Becquerel was wrong. In that case, it would be ridiculous to try to understand what led Becquerel to make mistakes (there would be no mistakes to be explain); instead, it would be relevant to understand why he accepted the claims of his critics and changed his ideas about

---

[16] Giora Hon (1989) has recently proposed a classification that does not include some of the crude errors commited by Becquerel.

uranium radiation. If, on the other hand, Becquerel commited a series of mistakes, it is relevant to understand the source of those mistakes, and discuss the existence of methodological rules that could help to avoid such errors.

11.1 *Reflection*

Consider, first, Becquerel's observation of specular reflection of uranium radiation. According to the accepted laws of geometrical optics, *no kind of radiation* could produce the effect described by Becquerel, even if it did suffer regular reflection at the surface of the metal mirror. There are a few possibilities of interpreting what happened:

a) Becquerel did make that experiment, and
    a1) he obtained a photograph exactly as he described it; or
    a2) he obtained a photograph with an irregular fogged circle where he thought there was a spot corresponding to the defect of the mirror; or
    a3) he obtained a photograph that showed no spot corresponding to the defect of the mirror, and misreported his observation.
b) Becquerel never made that experiment and described an imaginary effect.

In my opinion, the first case (a1) is unacceptable because it would conflict with our current knowledge. Becquerel never published the photograph of this experiment. If it existed, if it was as described by Becquerel, and if it were shown to me, I would be as surprised as if someone had shown to me a *perpetuum mobile*.

The second case (a2) is possible – it is the most benevolent interpretation of Becquerel's mistake. In that case, Becquerel was deceived by his theoretical expectation, and saw what he expected to see in a photograph where nothing or anything could be seen (as in a Rorschach spot). However, if Becquerel did believe that his photograph showed a spot corresponding to the defect of the mirror, why didn't he publish the photograph, as he did in other cases? It could have happened that, at first, Becquerel was misled by his photograph (a2), and afterwards, convinced that the effect did not exist, noticed that the photograph was not sufficiently sharp to support any claim, and chose not to publish it. In that case, he concealed his former mistake, but did not conterfeit his early report.

Alternatives (a3) and (b) would correspond to fraud.

11.2 *Refraction*

As described above, Becquerel found evidence for the refraction of his penetrating radiation in three different experiments: two with uranium nitrate (sealed tube and prism experiments) and one with calcium sulphide. The photographs corresponding to the uranium nitrate experiments have never been published. The photograph of the calcium sulphide experiment was published several times.

In the case of each uranium nitrate experiment, it is possible to distinguish several possibilities, as above:

a) Becquerel did make that experiment, and
    a1) he obtained a photograph exactly as he described it; or
    a2) he obtained a photograph with irregular fogged spots where he thought there was positive evidence for the effect he described; or
    a3) he obtained a photograph that showed no evidence for the effect he expected, and misreported his observation.
b) Becquerel never made that experiment and described an imaginary effect.

As in the case of reflection, case (a1) is unlikely because it would conflict with our current knowledge, and case (a2) is possible but problematic, because Becquerel never published those photographs.

In the experiment with calcium sulphide, however, the case is completely different. We know that Becquerel made the experiment and obtained a photograph exactly as he described it. Therefore, this experiment deserves a special analysis.

11.3 *Calcium sulphide radiation*

It is necessary to discuss, first, how could Becquerel observe the emission of penetrating radiation by calcium sulphide, given that we know that calcium sulphide emits no penetrating radiation. Then, it is necessary to discuss the evidence for reflection and refraction of radiation in that experiment.

There are several possible interpretations of Becquerel's observation of a penetrating radiation emitted by calcium sulphide:

a) Contrary to our current belief, calcium sulphide does indeed emit a penetrating radiation in some unknown circumstances.
b) Calcium sulphide emits no penetrating radiation, and Becquerel observed the radiation emitted by another substance:
   b1) there was an impurity mixed with calcium sulphide; or
   b2) Becquerel took a radioactive (uranium) sample for calcium sulphide.
c) The photographic plate was not adequately protected against light emitted from calcium sulphide; what Becquerel observed was light, not a penetrating radiation.

Interpretation (a) is unlikely, but at that time it could be supported by Niewenglowsky's experiments that had also detected penetrating radiation emitted by calcium sulphide. The impurity interpretation (b1) is unlikely because the effect observed by Becquerel was very strong (even stronger than those obtained with pure uranium compounds) and because it disappeared after some time. We know, of course, that there are some short-lived strongly radioactive substances, but they were not available in Becquerel's laboratory. Interpretation (b2) is very unlikely, given that Becquerel was well acquainted with luminescent substances, and calcium sulphide has a long-lived, strong phosphorescence of peculiar colour.

Interpretation (c) – a rude negligence – is the most likely interpretation, because the radiation emitted by calcium sulphide exhibited reflection and refraction. There is a difficulty, however: Becquerel described that in those experiments the photographic plate was wrapped in black paper and there was an aluminium plate, 2 mm thick, between the samples and the plate. Even if the paper was not opaque enough, light would never pass through that metal plate. Light could only affect the photographic plate *if the aluminium plate was missing*.

Could that be the case? It seems unbelievable that Becquerel could be so negligent. Such an error could perhaps be ascribed to Becquerel's laboratory assistant, called Louis Matou.

Is there any independent evidence that the aluminium plate was missing? Yes, there is. Contrary to his uranium photographs, that exhibited diffuse boundaries, Becquerel's calcium sulphide photograph is remarkably sharp. It shows details that could never be produced by any radiation, if the aluminium plate were between sample and photographic plate. Indeed: a point source of radiation can project a well defined shadow of an object at a large distance; however, an extended source of radiation (the phosphorescent sample inside the glass tube) will produce a poorly defined shadow. In the circumstances of Becquerel's experiment, if there were a distance of 2 mm between the thin glass plate that was used to close the tube and the photographic plate, the border of the glass plate could not be as sharp as it is. Scientifically speaking, the photograph published by Becquerel cannot have been produced under the circumstances he described.

Even if the metal plate was missing, it would be difficult for light emitted by calcium sulphide to pass through black paper. Hence, it is likely that the paper wrapping was also missing.

This interpretation of Becquerel's experiment elucidates other misterious aspects: Why was the radiation of calcium sulphide reflected and refracted by glass? Because the observed radiation was simply light. Why didn't the same sample of calcium sulphide emit penetrating radiation in later experiments? Because in later experiments Becquerel was careful enough to wrap the plate and/or to use an aluminium plate.

Only this interpretation (rude negligence) can fit our current physical knowledge and the effects described and registered by Becquerel. Notice that, in that case, Becquerel never suspected that he comitted such an error, since he was proud of his experiment, he exhibited and published his photograph, and he never concealed that evidence.

11.4 *Stimulation of emission of radiation by light*

As described above, Becquerel provided two classes of evidence for the excitation of uranium radiation by light. One kind was the comparison between the spots produced by two samples upon a photographic plate. The other one was the measurement of the speed of discharge of an electroscope.

In the case of the first method, only a very strong difference could be unabiguously detected. Indeed: the samples used by Becquerel were not exactly equal, but only roughly similar – and any observed small difference in radiation could be ascribed to a difference in the samples themselves. Besides that, visual comparison between two dark spots in a photographic plate is highly subjective, except if one is much darker than the other. Becquerel reported that when a flake of the uranium salt was illuminated by an electric arc or by discharge of a Leyden bottle, "the impressions are noticeably darker" (*les impressions sont notablement plus noires*) (Becquerel, 1896d, p. 691). "Noticeable", of course, can be interpreted either as striking or as merely observable.

According to our current physical knowledge, Becquerel could not have observed any strong increase in radiation emission, because uranium radiation is not excited by light. The electric arc could *heat* the uranium salt flake, and that could increase the photographic effect below the illuminated sample – but discharge of a Leyden bottle would not produce the same effect. It seems likely that the spots were very similar to one another, but Becquerel saw one of them darker than the other because he expected the effect to occur. The photographic evidence was never published by Becquerel. This case is comparable to Becquerel's mistaken experiments on reflection and refraction of uranium radiation.

On the other hand, Becquerel's electroscopic experiment showed a new kind of error. In this case, it is possible to detect Becquerel's error using information published seven years later. The table containing the 28th March 1896 measurements was published in Becquerel's book (Becquerel, 1903b, p. 20) and it allows us to recognize several problems: (a) lack of precision of measuments; (b) lack of reproductibility; (c) only a single series of measurements was made. Let us briefly describe those problems.

Becquerel measured time in hours and minutes, and measured angles in degrees and tenths of degree. Given the size of the electroscope leaves, it is likely that his instrument only allowed estimation of tenths of degree by visual interpolation[17]. Parallax and mistaken interpolation could easily introduce random errors of the order of half degree.

The inicial rate of discharge of the electroscope was computed from *a single pair of measurements*: at 1:44 hours the angle between the electroscope leafs was 10.3 degrees, and at 1:54 hours the angle was 4.5 degrees. Therefore, there was a motion corresponding to 5.8 degrees in 10 minutes, or 0.58 degrees/minute. Transforming to seconds of arc per second of time, one obtains 34.8"/s, with an error that could amount to about 10%.

In his paper of 1896, as described above, Becquerel presented a table that showed that at 1:45 p.m. the speed was 38.18"/s. Besides the computational error, Becquerel's figures published in 1896 seemed to imply a very large precision.

The second measurement (without any screen), between 6:12 and 6:25 p.m., gave a discharge speed of 0.54 degrees/minute or 32.4"/s (again, the number published in 1896 is wrong). And

---

[17] If the electroscope leaves had a length of 6 cm, a displacement of one millimetre would correspond to one degree. It is unlikely that the scale of the electroscope had divisions smaller than that.

finally, the next day, Becquerel measured four different speeds of discharge of the electrocope: 0.43, 0.50, 0.51 and 0.56 degrees/minute, with a mean speed of 0.53 degrees/minute or 31.8"/s. The difference between the first and second (or third) measurement is smaller than 10% and does not seem significative.

There is, however, a deeper problem with Becquerel's measurements. In the series of measurements of 29th March, as just shown above, the measured speeds varied between 0.43 and 0.56 degrees/minute. Why was there so large a variation? The reason is very simple: the speed of the electroscope leaves was not constant, but depended on the initial angle. It can be easily perceived from Becquerel's tables that the angular speed was smaller for larger initial angles. Becquerel, however, had wrongly stated that "when we follow the progressive approach of the electroscope golden leaves during discharge, it is recognized that, for apertures that do not exceed 30°, the angular changes are very sensibly proportional to time [...]" (Becquerel, 1896d, pp. 689-690).

Of course, in order to admit *quantitive comparison* between measurements, his experiments should always have been done between the same initial and final angles. *This was never the case*. In the first measurement described above, the angle varied from 10.3 to 4.5 degrees; in the second, from 13.8 to 6.8 degrees; and in the third, from 30.0 to 0.5 degrees. The third measurement cannot, of course, be compared with the first or second. If we select from the third measurement the data taken between 5:36 and 5:59 o'clock, when the angle varied from 13.5 to 0.5 degrees, the rate of discharge was 0.56 degrees/minute – that is, a value that is *between* the results of the first and second measurements.

What conclusion could be drawn? Of course, that no decrease of radiation intensity was detected. Becquerel drew, however, the opposite conclusion. It seems that Becquerel was not skilful in quantitative research. He overlooked elementary rules about measurement and interpretation of data.

11.5 *Polarization of uranium radiation*

Becquerel's evidence for polarization of uranium radiation was, in a sense, similar to that for reflection and refraction: he compared two spots on a photographic plate, in a single trial, and concluded that one of them was "considerably stronger" than the other. However, in this case, Becquerel published the corresponding photograph and the occurrence of fraud is excluded. The photograph cannot be as easily interpreted as Becquerel claimed: indeed, it exhibits a diffuse dark spot where it is very difficult to see any difference between the intensity of radiation transmited through parallel or crossed tourmalines. Becquerel was certainly misled by his theoretical expectations, as in the case of the photographic evidence for the increase of radiation by light stimulation.

11.6 *Summary*

In all cases, Becquerel was strongly influenced by his theoretical preconceptions. His mistakes, however, belong to different kinds.

a) Calcium sulphide: he (or his assistant) committed a rude mistake – the experiment was not performed as planned and described. Once this error was done and overlooked, the photographic evidence did support Becquerel's interpretation.

b) Electroscopic measurement of stimulation of radiation emission by light: Becquerel was unable to follow some well known rules about measurement and manipulation of quantitative data.

c) Photographic evidence for polarization and stimulation of radiation by light: the photographic evidence was inconclusive, but Becquerel arrived nevertheless to definite conclusions. The first case cannot be classified as fraud, as Becquerel published the corresponding photograph and claimed that it supported his interpretation.

d) Photographic evidence for reflection and refraction of uranium radiation: either the photographic evidence was inconclusive, as above, or Becquerel frauded the experiment.

12. COULD BECQUEREL'S ERRORS HAVE BEEN AVOIDED?

In a few years, Becquerel's mistakes were corrected by the scientific community. However, why should we trust the authors that reported that Becquerel was wrong, instead of Becquerel? Can we interpret the episode as a blind conflict between people with different preconceptions – each one seeing in his/her experiment what he/she expected to observe?

It seems impossible, in this episode, to adopt a relativist interpretation, since Becquerel was soon convinced that his earlier results were wrong. In some cases, he presented his early evidence and tried to convince the scientific community that he had been unavoidably misled by objective data, but he never claimed that the refutation of his claims was problematic or wrong.

Were Becquerel's mistakes unavoidable? Could a better experimental methodology lead Becquerel to correct results? Maybe. Had Becquerel been as careful and critical as later tradition said he was[18], he would *deserve* the name of "discoverer of radioactivity".

The possibility of mistakes in empirical investigation had been discussed since the beginning of the scientific revolution, by Francis Bacon and other authors. In the late 19th century, we find several works calling again the attention of scientists to the need for special care in experimental research. William Stanley Jevons, for instance, stresseed the role of preconceptions:[19]

---

[18] According to Oliver Lodge, "Henri Becquerel set himself carefully and critically to examine the kind of penetrating radiation which fluorescent substances exposed to light might possibly be found to emit [...]" (Lodge, 1912).

[19] The first edition of William Stanley Jevons' book *The principles of science* was published in 1874. I present citations of this book, not of more recent books on philosophy or methodology of science, to avoid anachronism.

Every observation must in a certain sense be true, for the observing and recording of an event is in itself an event. But before we proceed to deal with the supposed meaning of the record, and draw inferences concerning the course of nature, we must take care to ascertain that the character and feelings of the observer are not to a great extent the phenomenon recorded. The mind of man, as Fancis Bacon said, is like an uneven mirror, and does not reflect the events of nature without distortion. [...]

It is difficult to find persons who can with perfect fairness register facts for and against their own peculiar views. Among uncultivated observers the tendency to remark favourable and forget unfavourable events is so great, that no reliance can be placed upon their supposed observations. (Jevons, 1958, p. 402)

Accordingly, Jevons recommended:

Thus the successful investigator must combine diverse qualities: he must have clear notions of the result he expects and confidence in the truth of his theories, and yet he must have that candour and flexibility of mind which enable him to accept unfavourable results and abandon mistaken views. (Jevons, 1958, p. 404)

Of course, it is easier to talk about the correct attitude than to show how it can be attained. There is, however, a simple methodological rule: to repeat and to vary experiments.

Even when we are not aware by previous experience of the probable presence of a special disturbing agent, we ought not to assume the absence of unsuspected interference. If an experiment is of really high importance, so that any considerable branch of science rests upon it, we ought to try it again and again, in as varied conditions as possible. We should intentionally disturb the apparatus in various ways, so as if possible to hit by accident upon any weak point. (Jevons, 1958, p. 431)

Suppose Becquerel repeated and varied his experiments: he would probably find contradictory results. If he were not too stubborn, it is likely that he would have been able to correct his former mistakes. Or was he too obstinate?

In several cases (such as the polarization experiment), Becquerel made a single experiment and jumped to conclusions. In other cases, he did repeat and vary his experiments – but that did not lead him to correct his former view. Such was the case of his refraction and light excitation experiments. He also repeated his calcium sulphide experiments, noticed a contradictory result, but only concluded that the sample had lost its capacity of emitting penetrating radiation. It seems that Henri Becquerel was, indeed, too stubborn to correct his own views. His theoretical confidence in his interpretation of the radiation emitted by uranium compounds as an invisible phosphorescence was so strong that he acted exactly as Jevons described the "uncultivated observers" would do. No methodological rule will lead to safe results in the hands of the wrong scientist.

Becquerel's mistakes were avoidable and a better experimental methodology could have lead to correct results – in the hands of another scientist. Becquerel himself, however, did not have the adequate experimental training and correct attitude.

We must acknowledge that there are highly qualified observers, such as Newton and Faraday (Jevons, 1958, chapter xxvi, pp. 574-593) and unreliable observers (such as Becquerel and Blondot). One of the signs of a good experimenter is the capacity of finding and accepting evidence opposite to his own preferred ideas. Of course, even the best experimenter will commit mistakes. For that reason, it is unavoidable that the construction of science should be a collective enterprise.

13. FINAL COMMENTS

Henri Becquerel's experimental research on the phenomenon we now call "radioactivity" was full of serious mistakes. He ascribed to uranium radiation several properties – such as reflection, refraction, polarization, and increase by light stimulation – that were corrected by other researchers. In the study of uranium radiation – as in the case of N-rays – Henri Becquerel was a careless and perhaps unfair observer, misguided by his preconceptions. Later, however, he was socially successful in reinterpreting his early work and convincing the scientific community that his research was seldom mistaken, and that he had himself corrected his earlier mistakes.

ACKNOWLEDGEMENTS

This work was produced while the author was a visiting scholar of the Department of History and Philosophy of Science, University of Cambridge, and visiting fellow of Wolfson College (1995-1996). The author is grateful to the São Paulo State Research Foundation, Brazil (FAPESP) for supporting this research.

BIBLIOGRAPHIC REFERENCES

BADASH, Lawrence. The discovery of thorium's radioactivity. *Journal of Chemical Education* **43**: 219-220, 1966.

BECQUEREL, Edmond. Note sur la phosphorescence produite par insolation. *Annales de Chimie et de Physique* [3] **22**: 244-255, 1848.

BECQUEREL, Henri. Relations entre l'absorption de la lumière et l'émission de la phosphorescence dans les composés d'uranium. *Comptes Rendus Hebdomadaires des Séances de l'Académie des Sciences de Paris* **101**: 1252-1256, 1885.

BECQUEREL, Henri. Sur les différentes manifestations de la phosphorescence des minéraux sous l'influence de la lumière ou de la chaleur. *Comptes Rendus Hebdomadaires des Séances de l'Académie des Sciences de Paris* **112**: 557-563, 1891.

BECQUEREL, Henri. Sur les radiations émises par phosphorescence. *Comptes Rendus Hebdomadaires des*

Séances de l'Académie des Sciences de Paris **122**: 420-421, 1896 (a).

BECQUEREL, Henri. Sur les radiations invisibles émises par les corps phosphorescents. *Comptes Rendus Hebdomadaires des Séances de l'Académie des Sciences de Paris* **122**: 501-503, 1896 (b).

BECQUEREL, Henri. Sur quelquer propriétés nouvelles des radiations invisibles émises par divers corps phosphorescents. *Comptes Rendus Hebdomadaires des Séances de l'Académie des Sciences de Paris* **122**: 559-564, 1896 (c).

BECQUEREL, Henri. Sur les radiations invisibles émises par les sels d'uranium. *Comptes Rendus Hebdomadaires des Séances de l'Académie des Sciences de Paris* **122**: 689-694, 1896 (d).

BECQUEREL, Henri. Sur les propriétés différentes des radiations invisibles émises par les sels d'uranium, et du rayonnement de la paroi anticathodique d'un tube de Crookes. *Comptes Rendus Hebdomadaires des Séances de l'Académie des Sciences de Paris* **122**: 762-767, 1896 (e).

BECQUEREL, Henri. Émission de radiations nouvelles par l'uranium métallique. *Comptes Rendus Hebdomadaires des Séances de l'Académie des Sciences de Paris* **122**: 1086-1088, 1896 (f).

BECQUEREL, Henri. Note sur quelques propriétés du rayonnement de l'uranium et des corps radio-actifs. *Comptes Rendus Hebdomadaires des Séances de l'Académie des Sciences de Paris* **128**: 771-777, 1899.

BECQUEREL, Henri. Sur le rayonnement de l'uranium et sur diverses propriétés physiques du rayonnement des corps radio-actifs. Vol. 3, pp. 47-78, in: GUILLAUME, Charles-Édouard; POINCARÉ, Lucien (eds.). *Rapports Présentés au Congrès International de Physique réuni a Paris en 1900*. 3 vols. Paris: Gauthier-Villars, 1900.

BECQUEREL, Henri. Sur la radioactivité secondaire. *Comptes Rendus Hebdomadaires des Séances de l'Académie des Sciences de Paris* **132**: 734-739, 1901.

BECQUEREL, Henri. Sur la radio-activité de la matière. *Notice of the Proceedings of the Meetings of the Members of the Royal Institution of Great Britain* **17**: 85-94, 1902.

BECQUEREL, Henri. Recherches sur une propriété nouvelle de la matière – activité radiante spontanée ou radioactivité de la matière. *Mémoires de l'Académie des Sciences de l'Institut de France* **46**: 1-360, 1903 (a).

BECQUEREL, Henri. *Recherches sur une propriété nouvelle de la matière: activité radiante spontanée ou radioactivité de la matière.* Paris: Firmin-Didot, 1903 (b).

BECQUEREL, Henri. Sur une propriété nouvelle de la matière, la radio-activité. *Les Prix Nobel* **3**: 1-15, 1903 (c).

BECQUEREL, Henri. Enquête. Les rayons N existent-ils? Déclaration de M. H. Becquerel, membre de l'Institut, professeur de physique au Muséum et à l'École Polytechnique. *Revue Scientifique* [5] **2**: 718, 1904.

BECQUEREL, Jean. Sur l'émission simultanée des rayons N et N_1. *Comptes Rendus Hebdomadaires des Séances de l'Académie des Sciences de Paris* **138**: 1332-5, 1904.

BLONDOT, René. Sur une nouvelle espèce de lumière. *Comptes Rendus Hebdomadaires des Séances de l'Académie des Sciences de Paris* **136**: 735-738, 1903 (a).

BLONDOT, René. Sur de nouvelles sources de radiations susceptibles de traverser les métaux, le bois, etc., e sur de nouvelles actions produites par ces radiations. *Comptes Rendus Hebdomadaires des Séances de l'Académie des Sciences de Paris* **136**: 1227-1229, 1903 (b).

BLONDOT, René. Enquête. Les rayons N existent-ils? Déclaration de M. Blondot, Professeur à l'Université de Nancy. *Revue Scientifique* [5] **2**: 620-621, 1904.

BOUTY, E. Les rayons de Röntgen et la photographie à travers les corps opaques. *Revue Scientifique* [4] **5**: 610-617, 1896.

CURIE, Marie Sklodowska. Les rayons de Becquerel et le polonium. *Révue Générale des Sciences* **10**: 41-50, 1899.

ELSTER, Johann Philipp Ludwig Julius & GEITEL, Hans Friedrich Karl. Versuche über Hyperphosphorescenz. *Jahresbericht des Vereins für Naturwissenschaft zu Braunschweig* **10**: 149-153, 1897.

GALITZINE, B. & KARNOJITSKY, A. de. Recherches concernant les propriétés des rayons X. *Comptes Rendus Hebdomadaires des Séances de l'Académie des Sciences de Paris* **122**: 717-718, 1896.

GIESEL, Fritz. Über die Ablenkbarkeit der Becquerelstrahlen im magnetischen Felde. *Annalen der Physik und Chemie* [2] **69**: 834-836, 1899.

HENRY, Charles. Augmentation du rendement photographique des rayons Roentgen par le sulfure de zinc phosphorescent. *Comptes Rendus Hebdomadaires des Séances de l'Académie des Sciences de Paris* **122**: 312-4, 1896.

HOLMES, Frederic L. Do we understand historically how experimental knowledge is acquired? *History of Science* **30**: 119-136, 1992.

HON, Giora. Towards a typology of experimental errors: an epistemological view. *Studies in History and Philosophy of Science* **20**: 469-504, 1989.

JEVONS, William Stanley. *The principles of science: a treatise on logic and scientific method*. New York: Dover, 1958.

KELVIN, Lord; BEATTIE, J. C.; DE SMOLAN, M. S. Experiments on the electrical phenomena produced in gases by Röntgen rays, by ultraviolet light, and by uranium. *Proceedings of the Royal Society of Edinburgh* **21**: 393-428, 1897.

KLOTZ, Irving M. The N-ray affair. *Scientific American* **242** (5): 122-129, May 1980;

LAGEMANN, Robert T. New light on old rays: N rays. *American Journal of Physics* **45**: 281-284, 1977.

LE BON, Gustave. La lumière noire et les propriétés de certaines radiations du spectre. *Revue Scientifique* [4] **7**: 689-691, 1897.

LE BON, Gustave. La luminescence invisible. *Revue Scientifique* [4] **11**: 106-109, 1899.

LODGE, Oliver. The discovery of radioactivity, and its influence on the course of physical science (Becquerel memorial lecture). *Journal of the Chemical Society* **101**: 2005-2032, 1912.

MALLEY, Marjorie. The discovery of atomic transmutation: scientific styles and philosophies in France and Britain. *Isis* **70**: 213-223, 1979.

MALLEY, Marjorie. The discovery of the beta particle. *American Journal of Physics* **39**: 1454-1460, 1971.

MARTINS, Roberto de Andrade. Becquerel and the choice of uranium compounds. *Archive for History of Exact Sciences* **51** (1): 67-81, 1997.

MARTINS, Roberto de Andrade. *Os "raios N" de René Blondlot: uma anomalia na história da física*. Rio de Janeiro: Booklink; São Paulo, FAPESP; Campinas, GHTC, 2007.

NIEWENGLOWSKI, Gaston Henri. Sur la proprieté qu'ont les radiations émises par les corps phosphorescents, de traverser certains corps opaques à la lumière solaire, et sur les expériences de M. G. Le Bon, sur la lumière noire. *Comptes Rendus Hebdomadaires des Séances de l'Académie des Sciences de Paris* **122**: 385-386, 1896.

NYE, Mary Jo. N-rays: an episode in the history and psychology of science. *Historical Studies in the Physical Sciences* **11**: 125-156, 1980.

PICKSTONE, John V. Past and present knowledges in the practice of the history of science. *History of Science* **33**: 203-224, 1995.

PIÉRON, Henri. Grandeur et décadence des rayons N. Histoire d'une croyance. *L'Année Psychologique* **13**: 143-169, 1907.

ROSMORDUC, Jean. Une erreur scientifique au début du siècle: "les rayons N". *Revue d'Histoire des Sciences et de leurs Applications* **25**: 13-25, 1972.

RUTHERFORD, Ernest. Uranium radiation and the electrical conduction produced by it. *London, Edinburgh and Dublin Philosophical Magazine and Journal of Science* [5] **47**: 109-163, 1899.

RUTHERFORD, Ernest. *Radioactivity*. Cambridge: Cambridge University Press, 1904.

RUTHERFORD, Ernest & SODDY, Frederick. The radioactivity of thorium compounds. I. An investigation of the radioactive emanation. II. The cause and nature of radioactivity. *Journal of the Chemical Society, Transactions* **81**: 321-350, 837-860, 1902 (a).

RUTHERFORD, Ernest & SODDY, Frederick. The cause and nature of radioactivity. *Philosophical Magazine (series 6)* **4**: 370-396, 569-585, 1902 (b).

RUTHERFORD, Ernest & SODDY, Frederick. Radioactive change. *Philosophical Magazine (series 6)* **5**: 561-576, 576-591, 1903.

SCHMIDT, Gerhard C. Ueber die von den Thorverbindungen und einigen anderen Substanzen ausgehende Strahlung. *Annalen der Physik und Chemie* [2] **65**: 141-151, 1898.

STRADLING, George Flowers. A resumé of the literature of the N rays, the N_1 rays, the physiological rays and the heavy emission, with a bibliography. *Journal of the Franklin Institute* **164**: 57-74, 113-130, 177-199, 1907.

THOMPSON, Silvanus P. On hyperphosphorescence. *Report of the 66th Meeting of the British Association for the Advancement of Science* [**66**]: 713, 1896 (a).

THOMPSON, Silvanus P. On hyperphosphorescence. *The London, Edinburgh and Dublin Philosophical Magazine and Journal of Science* [5] **42**: 103-107, 1896 (b).

THOMSON, J. J. The Röntgen rays. *Nature* **53**: 581-583, 1896.

TÖRNEBLADH, H. R. Physics 1903. Presentation speech by Dr. H. R. Törnebladh, President of the Royal Swedish

Academy of Sciences. Pp. 47-51, in: *Nobel lectures including presentation speeches and laureates' biographies. Physics. 1901-1921.* Amsterdam: Elsevier, 1967.

TRENN, Thaddeus. J. *The self-splitting atom: the history of the Rutherford-Soddy collaboration.* London: Taylor and Francis, 1975.

TROOST, Louis Joseph. Observation à l'occasion de la communication de M. H. Becquerel. *Comptes Rendus Hebdomadaires des Séances de l'Académie des Sciences de Paris* **122**: 694, 1896.

THE GUIDING HYPOTHESIS OF THE CURIES' RADIOACTIVITY RESEARCH: SECONDARY X-RAYS AND THE SAGNAC CONNECTION

Roberto de Andrade Martins

Abstract: Pierre and Marie Curie's main discoveries on radioactivity are usually regarded as empirical investigations that were developed without any theoretical guidance. Their papers avoid indeed theoretical discussion, but it is possible to identify the main hypothesis that directed their work. They thought that the radiation emitted by uranium compounds (and, later, by other similar substances) was similar to the secondary radiation emitted by heavy metals when they are hit by X-rays. This hypothesis, together with other relevant assumptions, was suggested by Georges Sagnac's investigation on X-rays. This paper describes Sagnac's studies and how the acceptance of the secondary radiation hypothesis guided the study of radioactivity by the Curies. For the Curies, this hypothesis explained one of the anomalous characteristics of radioactivity – the continuous emission of energy without any noticeable change of the emitting bodies. When the magnetic deviation of the beta-rays of radioactive bodies was discovered, in 1899, this presented a challenge to their hypothesis. They carefully checked that discovery, and attempted to produce a magnetic deflection of X-rays, with negative results. However, in 1900 Pierre Curie and Georges Sagnac investigated secondary X-rays and concluded that they contained both "soft" X-rays and a negatively charged radiation (similar to beta-rays). Because of those results, they still kept their faith in the secondary radiation hypothesis at the time when Rutherford and Soddy began to develop the disintegration theory of radioactivity.
Keywords: radioactivity; X-rays; history of physics; Curie, Marie; Curie, Pierre; Sagnac, Georges

MARTINS, Roberto de Andrade. *Historical Essays on Radioactivity*. Extrema: Quamcumque Editum, 2021.

167

1. INTRODUCTION

Pierre and Marie Sklodowska Curie's main discoveries on radioactivity are usually regarded as empirical investigations that were developed without any theoretical guidance. Their approach has been contrasted to Ernest Rutherford's, and it has been suggested that the use of concrete models and hypotheses by the later contributed to his success, where the Curies failed.

Up to 1900, the French were the leaders in the study of radioactivity (Jauncey, 1946). However, the understanding of radioactivity as a phenomenon of atomic transmutation came from abroad. How did they loose their leadership?

In 1899 the Curies discovered that an object placed near to a strongly radioactive source became radioactive. Ernest Rutherford also noticed that bodies near thorium became radioactive. In both cases, it was noticed that the radioactivity of those bodies was short-lived. The Curies described the phenomenon as an "induced activity", and they initially rejected Rutherford's proposal that it could be produced by a material emanation coming from the radioactive substances. Rutherford's approach led to the discovery of radon and of atomic transmutation. The Curies' approach to induced radioactivity led to a mere accumulation of facts and attempts to discuss them in a more general, abstract way.

According to some historians, the Curies systematically adhered to an abstract and timid approach to radioactivity, attempting to produce generalisations from observed facts and following a thermodynamic perspective. Rutherford, on the other hand, is described as a bold researcher who framed concrete, risky hypotheses and allowed them to guide his research.

The difference between the attitudes of Rutherford and the Curies has been sometimes described as due to contrasting personalities; or to national differences (see Malley, 1979; Nye, 1993, for a discussion of the French and English national styles)[1]; or to the distinct research schools to which they belonged (Davis, 1995).

However, before attempting to *explain* a fact, it is wise to check whether the fact is true, or an artefact produced by the historian's analysis.

I maintain in this paper that the attitudes of the Curies and Rutherford respecting the use of hypotheses were not widely different as has been claimed.

2. HYPOTHESES IN THE RESEARCHES OF BECQUEREL AND THE CURIES

It is usually assumed that Henri Becquerel's research was also purely empirical. In a former paper I have argued that Becquerel's work was guided by a hidden hypothesis concerning the violation of Stokes' law in uranium and its compounds (Martins, 1997). I contend that, in a similar way, the Curies' researches on radioactivity were strongly directed by a hypothesis – one that was not as concealed or secret as in the case of Becquerel's work. Indeed, the Curies' papers usually averted theoretical discussion and presented no hint of a guiding hypothesis. However, in other papers it is possible to identify plain clues of the main hypothesis that directed their work.

The hypothesis that will be discussed here appeared in print, for the first time, in Marie Sklodowska Curie's paper announcing that thorium emitted a penetrating radiation, just like uranium. She suggested that the radiation emitted by uranium and thorium compounds (and, later, by other similar substances) was produced by an unknown radiation coming from space, that was transformed inside those substances, in the same way as X-rays can be transformed into secondary rays. This hypothesis, together with other relevant assumptions, was suggested by Georges Sagnac's investigation on X-rays.

---

[1] Of course, national differences between England and France are difficult to apply in this specific case, because Rutherford was from New Zealand and Marie Sklodowska Curie was Polish.

Analogy with the secondary rays of the Röntgen rays. – The properties of the rays emitted by uranium and thorium are very similar to those of the secondary rays of the Röntgen rays, recently studied by Mr. Sagnac. Besides that, I have noticed that under the action of the Röntgen rays, uranium, pitchblende and thorium oxide emit secondary rays which, from the point of view of the discharge of electrified bodies, often produce stronger effects than the secondary rays of lead. Among the metals studied by Mr. Sagnac, uranium and thorium would be placed in the neighbourhood of lead, and beyond it.

To elucidate the spontaneous radiation of uranium and thorium we could imagine that the entire space is always crossed by rays analogous to the Röntgen rays, but much more penetrating and that could only be absorbed by certain elements with a large atomic weight, such as uranium and thorium. (Sklodowska-Curie, 1898a, p. 1103)

Let us first make clear the meaning of Marie Sklodowska Curie's hypothesis. The starting point of her research was, of course, Henri Becquerel's investigation of the rays emitted by uranium and its compounds, in 1896-1897. Becquerel believed that those rays were similar to X-rays (or Röntgen rays). Although the nature of X-rays was not established at that time, Becquerel believed that they were high-frequency electromagnetic waves (beyond the ultraviolet). He supposed that uranium and its compounds could transform visible light into X-rays by a special phenomenon of phosphorescence violating Stokes's law. Led by his belief, [2] Becquerel reported observations to the effect that the radiation emitted by uranium compounds decreased slowly in the darkness, and increased after they were strongly illuminated; that the radiation of uranium compounds could be reflected by a metallic mirror, could be refracted by glass and polarised by a tourmaline

---

[2] See MARTINS, Roberto de Andrade. Becquerel's experimental mistakes, in this volume.

crystal. All his experiments seemed to confirm that uranium radiation was a high-frequency electromagnetic radiation.

Becquerel's early investigations on uranium radiation lasted from 1896 to 1897. During this period, there were very few other scientists who published any paper on the subject. The limited literature on this theme was one of the reasons that led Marie Sklodowska Curie to choose it as a research object for her PhD thesis. The decision was made towards the end of 1897. Her experimental researches started on the 16th of December, 1897 (Joliot-Curie, 1955, p. 106).

Georges Sagnac (1869-1928), a close friend of the Curies at that time, was one of the very few people who had carefully studied Becquerel's work before 1898 and he published a review paper on that phenomenon (Sagnac, 1896). It is possible that Sagnac influenced Marie Sklodowska Curie's choice of uranium radiation as a subject of research.

When Marie Curie started her work on uranium, both Georges Sagnac and Jean Perrin (1870-1942) – another friend of the Curies – were working on their PhD theses on X-rays. Perrin studied the discharge of electricity produced by X-rays. Sagnac studied the secondary radiation emitted by metals hit by X-rays. It is likely that Perrin and Sagnac discussed their researches with the Curies.

Becquerel had shown that the uranium rays were also able to discharge electrified bodies, as X-rays did. Marie Sklodowska Curie's first experiments, as shown in her laboratory notebook, were aimed at the study of the conductivity of air produced by uranium radiation. It is likely that she initially accepted all the conclusions published by Becquerel, and that she intended to develop a research similar to that of Jean Perrin, making a detailed study of all circumstances involved in the production of electric conduction by the uranium rays. Indeed, if the uranium rays were similar to X-rays, it was natural to use the researches on X-ray of her friends as a model for her own investigation. This circumstance could be the motivation for the specific choice made by Marie Curie at the beginning of her research.

Some early experiments led Marie Sklodowska Curie to conclude (as Becquerel had already noticed) that chemical reactions or temperature changes do not modify the intensity of the radiation emitted by uranium compounds (Joliot-Curie, 1955, pp. 106-108). The emission of the radiation only depended on the amount of uranium in a sample. Subsequently Curie noticed that all thorium compounds also emitted a similar radiation. As the emission was not influenced by external changes, it seemed an *atomic property* – and, of course, at that time, it was customary to regard atoms as unchangeable particles[3].

This was one of the explicit hypotheses presented by the Curies. It is well known that this hypothesis – that the emission of radiation was an atomic property – guided their successful search for new elements in pitchblende. The *atomic property hypothesis* was also confirmed when Marie Sklodowska Curie noticed that the amount of radiation emitted by uranium compounds is approximately proportional to their uranium contents, independently of the presence of other non-active elements in the substance.

Those facts did not conflict with Becquerel's initial conclusions. However, one of her early findings was that the radiation emitted by uranium and its compounds, carefully measured with an ionisation chamber, did not decrease in darkness and did not increase under strong illumination (Joliot-Curie, 1955, p. 106). Therefore, it did not behave as a phosphorescence phenomenon, as was supposed by Becquerel.

This discovery commanded a reflection on the source of energy behind the radiation phenomenon. Of course, for Becquerel the problem did not exist – the uranium radiation was

---

[3] Some years later, Frederick Soddy remarked: "The view that radioactivity is an atomic property necessitates, on the older view of the unchangeability of the atom, that the activity should be in all cases a permanent property of the matter exhibiting it." (Soddy, 1905, p. 256)

just a form of energy that had been absorbed by the uranium compounds from light, and was slowly released under the form of penetrating radiation. However, since that interpretation was not correct, it became imperative to find out the energy source behind the emission of radiation by uranium and thorium. This was probably the motive that led the Curies to formulate their second hypothesis (the penetrating radiation hypothesis), that has already been pointed out.

On April 12, Marie Sklodowska Curie's first paper on the radiation of thorium was read by Gabriel Lippman at the French Academy of Science. In a period of less than 4 months, besides obtaining several relevant experimental results, the Curies had also framed the hypotheses that would guide their future research, abandoning Becquerel's perspective concerning the uranium phenomenon[4].

3. SAGNAC'S INFLUENCE

Marie Sklodowska Curie's initial experiments were probably guided by Becquerel's ideas and by her own experimental results. When did Georges Sagnac's influence start?

This happened probably in the second half of March. The laboratory notebooks of Maric and Pierre Curie show that on the 16th of March most of the measurements required by the thorium paper had already been completed (Joliot-Curie, 1955, p. 109). Pierre was beginning to help Marie, and on that day they both wrote a summary of the previous work, probably as a draft for a future paper. They were probably excited with the new results, and it is likely that they would discuss their research with Jean Perrin and Georges Sagnac.

---

[4] One may wonder why it was Gabriel Lippman, not Henri Becquerel, who was asked by the Curies to report Marie's first paper to the Paris Academy of Science. Perhaps the reason was just that Marie had already worked with Lippman for some time, while studying the magnetism of several alloys. However, there might be another reason: the disagreement between Marie's results and Becquerel's ideas.

Sagnac was studying the secondary radiation produced by X-rays when they strike metals. Several researchers had attempted to detect the reflection of X-rays by metals and had failed. However, in some cases a dispersed radiation was observed coming from metals hit by X-rays. The initial interpretation was that the X-rays had been diffused or scattered by the metal; however, the diffuse radiation was less penetrating than the original one. Therefore, the metal had *transformed* the incident radiation. The phenomenon was similar to visible light fluorescence: the light emitted by a fluorescent substance has a smaller frequency than the incident radiation, according to Stokes' law. If the penetration of X-rays was related to their high frequency, then a secondary radiation of lower frequency was expected to be less penetrating.

In a paper where he described several properties of X-rays, including the production of secondary radiation, Sagnac remarked the similarity between the Röntgen rays and Becquerel's rays:

> It is appropriate to remind here the discovery due to H. Becquerel of new invisible radiations emitted during several months, without noticeable weakening, by uranium salts and especially by uranium, that have always been kept in darkness. Up to the present day it seems that there is no limit for the duration of those phenomena, for which S.-P. Thompson proposed the name *hyperphosphorescence*. We ignore if here there is really a transformation of radiations or simply a spontaneous radiation due to a new mechanism. Anyhow, those remarkable *uranium rays* are very close to the X-rays by their electrical properties. (Sagnac, 1898, p. 314)

The production of secondary radiation (or S-rays, as Sagnac called them) was especially strong when X-rays stroke metals of high atomic weight, such as lead. In the case of low atomic weight metals, such as aluminium, the incident rays traversed the metal without producing noticeable secondary radiation.

The secondary radiation was less penetrating than the original X-rays. For that reason, it was strongly absorbed and produced stronger effects (ionisation and photographic effects). The most penetrating X-rays passed by matter without noticeable energy loss, and therefore produced weak effects. The secondary radiation produced stronger effects, because its energy was easily absorbed by matter.

It is likely that Sagnac and the Curies discussed their mutual researches in the early months of 1898. Sagnac had been studying the secondary rays for some months, and several of his results had already been published, but he was continuing his researches during this period. The comparison between the two lines of research exhibited remarkable similarities. Marie Curie noticed that the rays emitted by uranium and thorium were similar to Sagnac's secondary rays:

1. Both the secondary rays and the uranium radiation were less penetrating than X-rays.
2. Only high atomic weight elements produced a large amount of easily absorbed secondary rays. The two elements that were known to emit Becquerel rays (uranium and thorium) were the elements with the highest atomic weight known at that time.

In her search for other substances that could emit penetrating rays, Marie Sklodowska Curie had noticed that some other elements (cerium, niobium, and tantalum) also seemed slightly active, but only uranium and thorium were very active. She commented:

> It is remarkable that the two more active elements, uranium and thorium, are those that have the highest atomic weights. (Sklodowska-Curie, 1898a, p. 1102)

This striking similarity suggested either to Sagnac or to the Curies the hypothesis of a penetrating radiation that could account for the energy emitted by uranium and thorium. Inasmuch as Marie Sklodowska Curie had already concluded that the emission of radiation by uranium was not similar to phosphorescence, and since the energy output seemed constant, the energy source could not be in the active material itself. It should come from outside, and the active substances just *transformed* some other form of energy existing in the environment into the Becquerel rays. The phenomenon could be analogous to the production of Sagnac's secondary rays by X-rays.

4. THE HYPOTHESIS OF A PENETRATING RADIATION

Marie Curie conjectured that a very penetrating unknown radiation existed everywhere. It produced no observable effects in ordinary matter but its transformation by heavy atomic weight elements could produce a detectable secondary radiation – the Becquerel rays.

This trend of ideas is not explicit in the early papers published by Marie Sklodowska Curie, but that seems a plausible reconstruction of the reasoning that led to the hypothesis of the penetrating radiation.

It seems that the hypothesis was not due to Sagnac. Indeed, in a paper on X-rays and secondary rays he published in 1898, Sagnac referred to the similarity between X-rays and the Becquerel rays, but did not compare them to the secondary rays. Also, as will be seen later, in 1901 this hypothesis was clearly ascribed to Marie Curie.

On the 1st April, the laboratory notebook shows that the Curies had already began to study the penetrating radiation conjecture. A series of experiments begun on this day, having the title "Effect of X-rays", attempted to detect changes in the amount of radiation emitted by uranium and other active materials when they were submitted to X-rays. The content of the notebook was described by Irène Joliot-Curie in the following way:

The experimental conditions are not precisely described. It seems that the idea was the following: the active matter was irradiated through the support, that absorbed little; the active matter was covered by a plate that could absorb only part of its radiation, but almost completely the X rays (this plate could be made of lead). They searched whether the X-rays excited or not a radiation analogous to the normal activity of the active substances. The active materials used were uranium, uranium oxide, orangite and pitchblende. (Joliot-Curie, 1955, p. 111)

It is obvious that, at this point, the relation between the secondary radiation produced by X-rays and the emission of Becquerel rays by uranium and other active substances was already at work, guiding the experiments of the Curies.

On the same day, the Curies compared the penetrating powers of the rays emitted by thorium and uranium. They observed that the radiation emitted by uranium was less penetrating than that emitted by thorium. In the case of secondary rays, those emitted by elements with higher atomic weight were also less penetrating. Therefore, this experiment disclosed another important similarity between the radiation of uranium and thorium and Sagnac's S-rays.

As was already described, a few days later Marie Sklodowska Curie's first paper was read by Gabriel Lippman. It contained a clear presentation of the penetrating radiation hypothesis. No alternative hypothesis was discussed in that paper. This circumstance strongly suggests that the Curies were immediately convinced that this was a correct assumption.

The atomic property hypothesis and the penetrating radiation hypothesis were in mutual agreement and reinforced each other. If the Becquerel rays were the outcome of the transformation of a penetrating radiation by elements of high atomic weight, this should be a property that depended on the properties of the *atoms* (not molecules), and the total amount of radiation produced in uranium compounds should only depend on the amount of the active element in the substance.

5. A RESEARCH GUIDED BY HYPOTHESES

However, there were two empirical exceptions to the atomic property hypothesis: pitchblende and chalcolite, two uranium minerals, were more active than metallic uranium. If the atomic property hypothesis were a mere empirical generalisation, it should have been rejected because of those exceptions. However, the Curies chose to retain this hypothesis and added another supposition: that there was another, unknown active element, in pitchblende. This risky supposition was already presented in Marie Sklodowska Curie's first paper:

> Two uranium minerals, pitchblende (uranium oxide) and chalcolite (phosphate of copper and uranium) are much more active than uranium itself. This is a very remarkable fact and it leads to the belief that those minerals can contain an element that is much more active than uranium. (Sklodowska-Curie, 1898a, p. 1102)

The strong confidence shown by the Curies in the atomic property hypothesis at this early stage of their researches is a strong evidence that this hypothesis was not just an empirical generalisation. It was part of a broader theoretical interpretation of the phenomenon, reinforced by Sagnac's work on the secondary radiation. Everything seemed to fit those hypotheses, and guided by those hypotheses the Curies embarked into a strenuous search for the unknown active element in pitchblende. The hypothesis of the penetrating radiation, and the hypothesis that radioactivity was an atomic phenomenon (but without any assumption of atomic change) guided those investigations of the Curies from April 1898 onwards.

The two hypotheses led them to the discovery, in 1898, of two new radioactive elements: polonium and radium. In their following papers describing the discovery of polonium and radium, the Curies did not mention the penetrating radiation hypothesis, but they did refer to the atomic property hypothesis.

6. TESTING THE HYPOTHESIS OF THE PENETRATING RADIATION

It seems that the search for the new active elements absorbed most of their time, and they did not attempt to check the penetrating radiation hypothesis. Meanwhile, other researchers did it. In September 1898 Johann Elster and Hans Geitel submitted to the journal *Annalen der Physik und Chemie* a paper where they discussed several contrasting explanations of the Becquerel rays – including Marie Sklodowska Curie's penetrating radiation hypothesis.

After a theoretical discussion of the several suggestions, Elster and Geitel described an experimental test of Marie Sklodowska Curie's conjecture (Elster & Geitel, 1898). The hypothetical penetrating radiation should be able to penetrate the whole atmosphere (equivalent to about 10 meters of water), the walls of laboratory buildings and metallic apparatus used in radiation experiments, without noticeable absorption. However it would be extravagant to suppose that it could penetrate any thickness of matter without suffering absorption. If radioactivity was produced by a penetrating radiation coming from space, it should be weaker in deep pits. Hence, they were led to test whether the emission of radiation by uranium suffered any change when it was observed in a very profound pit, about 850 metres deep. The experiment showed, however, that the activity of the radioactive sample was the same at the depth of 850 metres and at the ground level. The authors concluded:

> From those researches it seems to us that the hypothesis of production of Becquerel rays by other rays pre-existent in space is improbable to the highest degree. (Elster & Geitel, 1898, p. 740)

Marie Sklodowska Curie became aware of this paper soon after it publication, in December 1898, and referred to its negative result in a long review article she published in January 1899 (Sklodowska-Curie, 1899a, p. 50). In that paper, Marie presented for the first time *several* explanations that had been suggested for radioactivity – including the penetrating radiation hypothesis.

The Curies acknowledged that the result of the experiment made by Elster and Geitel presented a difficulty for the penetrating radiation conjecture. However, they did not give up their hypothesis. They possibly thought that the radiation was not noticeably absorbed by the materials constituting the crust of the Earth, for depths of a few hundred metres, because the minerals that build up that crust do not contain a strong proportion of high atomic weight elements. They devised another test, which was shortly described in Marie Sklodowska Curie's thesis. The date of this experiment is unknown:

> We have measured the radioactivity of uranium at noon and at midnight, thinking that if the Sun were the source of the hypothetical primary radiation, this could be partially absorbed in passing across the Earth. Experience did not provide any difference between the two measurements. (7, 1903, p. 140)

Although 850 metres of rock did not produce any change, the whole Earth should produce a noticeable absorption. If the penetrating radiation came from the Sun, the activity of uranium should be greater at noon than at midnight. No difference was observed, however.

Notice that the Curies did not gave up the penetrating radiation hypothesis after Elster and Geitel's results. Notice also that their own experiment could only possibly *confirm* the penetrating radiation hypothesis, because the negative outcome could be interpreted in a very simple way: the penetrating radiation did not come from the Sun.

7. INDUCED RADIOACTIVITY

The penetrating radiation hypothesis had a strong influence on the interpretation of the Curies concerning "induced radioactivity". They described their discovery of the phenomenon in the following manner:

> While studying the properties of strongly radioactive matter, prepared by us (polonium and radium), we have noticed that the rays emitted by those substances, acting upon inert substances, can communicate radioactivity to them, and that this radioactivity remains during a very long time. (Curie & Sklodowska-Curie, 1899, p. 714)

Notice that in the very description of the discovery, the Curies assumed that *the rays* had induced radioactivity in other materials. A "neutral" description of the phenomenon would only specify that an inert body put close to a strongly radioactive source would become radioactive.

After describing the experiments that they made concerning the phenomenon, the Curies concluded:

> The phenomenon of induced radioactivity is a type of secondary radiation due to the Becquerel rays. However, this phenomenon is different from the one that is known for Röntgen rays. Indeed, the secondary rays of the Röntgen rays that have been studies up to now are born immediately when the bodies that emit them are hit by the Röntgen rays and cease immediately with the suppression of the later. (Curie & Sklodowska-Curie, 1899, pp. 715-716)

Therefore, the hypothesis of the penetrating radiation and secondary rays was the basis of their initial interpretation of "induced radioactivity".

8. NON-ELECTROMAGNETIC RADIATIONS

In 1899, new advances brought fresh difficulties for the interpretation of radioactivity. When the Curies began their studies on uranium and its radiation, nobody suspected that those rays could be classified into several different types. They seemed very similar to soft X-rays. The situation changed in

1899. Ernest Rutherford studied the absorption of radiation by thin metallic foils and distinguished the α and β rays. In the same year, Fritz Giesel, Stefan Meyer and Egon von Schweidler noticed that some of those rays could be deviated by a magnetic field. Now, the similarity between the Becquerel rays and X-rays began to dwindle, and this was a challenge to the views embraced by the Curies.

The possibility of diverting the rays was first confirmed by Becquerel, and Pierre Curie himself soon confirmed that some of the rays produced by radium and polonium could also be deviated by a magnetic field. Was this a clear proof that they were charged particles? Perhaps it was not. The Curies decided to check this point. They soon described an experiment where they separated and collected the magnetically deflected rays (Rutherford's β rays). They were able to detect that those rays carried a negative electric charge (Curie & Sklodowska-Curie, 1900b). They seemed of the same nature as cathode rays. This finding threatened all their theoretical assumptions, because now the Becquerel rays could not be anymore assumed to be similar to the secondary radiation of X-rays.

The analogy could be maintained, however, if the X-rays also carried an electrical charge. The Curies tested this possibility, and did not find any clear evidence that X-rays conveyed electrical charges (Curie & Sklodowska-Curie, 1900b, p. 650).

Of course, they must have discussed the uncomfortable situation with Sagnac, and their old friend came to their rescue. Indeed, in 1898 Sagnac had noticed that the secondary rays contained, besides neutral radiation, some electrically charged particles. The evidence he obtained in 1898 was not altogether clear and he decided not to publish his discovery. However, in order to be able to claim priority afterwards, he placed a description of his research in a sealed envelope ("pli cacheté"), that was delivered to the French Academy of Sciences on July 18, 1898. In February 1900 he asked the Academy to open the envelope. Its content was then read and published (Sagnac, 1900).

That was a very important point. Pierre Curie and Georges Sagnac soon began a detailed joint investigation of this topic[5]. On April 9, 1900, they presented to the Paris Academy of Sciences the result of their research (Curie & Sagnac, 1900). They confirmed the previous result of the Curies that Röntgen rays do not carry a noticeable electric charge; however, "on the contrary, *the secondary rays* originating from the transformation of Röntgen rays *do convey electrical charges with them*, similar to cathode rays, as do the rays from radium" (Curie & Sagnac, 1900, p. 1013; emphasis of the authors).

The paper published by Curie and Sagnac did not mention the penetrating radiation hypothesis of radioactivity. However, the connection between the experiments and the hypothesis was made clear in another work on the same subject that they presented on the 3rd of May 1901 to the French Physical Society.

> The weak penetration power of the secondary rays of heavy metals reminds us Lenard's cathode rays: they can only reach a few centimetres in the atmospheric air, where they are strongly diffused. This analogy led us to search whether the secondary rays, which are strongly absorbed by the air, carry with them negative electric charges, since this is the fundamental characteristic of the cathode rays. The deviation of the rays by a magnetic or electric field will be the probable consequence of their electrification. *There is no contradiction between this hypothesis and those that have been developed by one of us*, since the beam spontaneously emitted by the *radium* of Mr. and Mrs. Curie is a mixture of rays with negative electricity, analogous to the cathode rays, that can be deviated by the magnetic field and by the electric field, together with rays that cannot be deflected, analogous to X-rays, which seem devoid of electrical charges. (Curie & Sagnac, 1902, p. 13; my emphasis)

---

[5] Let us remark that this was the only joint research ever done by Curie and Sagnac.

The paper did not elucidate what the authors meant by the hypothesis that had been developed by one of them. Was that hypothesis proposed by Sagnac, or by Pierre Curie? An anonymous account of the meeting of the French Physical Society where they presented this paper leaves no doubt concerning this point: "The existence of electrified secondary rays producing a deflectable beam is in accordance with the analogy between the secondary rays and the spontaneous rays of radioactive bodies pointed out by Mrs. Curie" (Anonymous, 1901, p. 499). Therefore, it is unlikely that Sagnac had suggested the penetrating radiation hypothesis. The two previous citations imply that it had been proposed by one of the Curies.

Pierre Curie and Georges Sagnac concluded from their experiments that the penetrating radiation hypothesis could be maintained in face of the new discovered properties of radiation. They noticed that the emission of negative charges together with the secondary rays was especially noticed in heavy metals – a circumstance that enhanced the similarity between this phenomenon and radioactivity (Curie & Sagnac, 1900; Curie & Sagnac, 1902).

9. THE FATE OF THE HYPOTHESIS OF SECONDARY RAYS

In 1900 the Curies presented a report on radioactivity to the International Congress of Physics that occurred in Paris. At the end of that report they discussed the nature of the Becquerel rays. They reported that those rays contain both charged rays, similar to the cathodic rays, and others that were similar to X-rays. The occurrence of both kinds of rays seemed easy to explain:

> This mixture should not amaze us. In the vacuum tubes the X-rays are born at the walls hit by cathodic rays. On the other side, when X-rays hit the bodies they produce the birth of the secondary rays studied by Mr. Sagnac, and those secondary

rays seem also to be formed by a mixture of non-deflectable rays and rays charged with electricity, analogous to cathode rays. There is therefore a strong analogy between the spontaneous emission of the radioactive bodies and the secondary rays of the Röntgen rays. This analogy had hit us since the beginning of this study, and afterwards it always became stronger.

[...]

According to what has just been said, it is possible to regard the Becquerel rays as a secondary emission due to some rays analogous to X-rays that traverse all space and every body.

If the emission in its totality is not a secondary emission, this could however be true for one of the two groups of rays; one could consider as primary rays either the non-deflectable rays, of the deflectable rays. (Curie & Sklodowska-Curie, 1900a, pp. 113-114).

The Curies also mentioned, at the end of their paper, the idea of a changing atom, but ascribed this idea to William Crookes and J. J. Thomson – not to themselves. It is plain that at that time the Curies had a strong confidence in the penetrating radiation hypothesis, and thought that it would remain acceptable at least for one of the types of radiation emitted by radioactive bodies.

It is possible to find other evidences that from 1900 to 1903 the Curies still accepted this hypothesis, notwithstanding the new facts that were being discovered. In 1903, for instance, Pierre Curie and André Laborde published the first measurement of the energy released by a radium salt. They concluded that 1 g of radium liberates about 100 calories per hour. The authors discussed the hypothesis that the energy liberation was due to an atomic change, and then they remarked: "The hypothesis of a continuous change of the atom is not the only one compatible with the release of heat by radium. This heat release can also be explained by supposing that the uranium makes use of an external energy of unknown nature." (Curie & Laborde, 1903, p. 675)

This suggests that Pierre Curie had not given up the penetrating radiation hypothesis, at this time. It is also relevant to notice that when Becquerel and the Curies received the Nobel Prize for their researches, in 1903, the former researcher maintained that the penetrating radiation hypothesis was still acceptable – although he preferred the idea of atomic transformation:

> Among the hypotheses which suggest themselves to fill the gaps left by current experiments, one of the most likely lies in supposing that the emission of energy is the result of a slow transformation of the atoms of the radioactive substances. [...]
>
> In this scheme, there would still be scope to wonder whether the transformation of the atom comprises a slow, spontaneous evolution, or whether it is the result of the absorption of external radiation beyond the range of our senses. If such a radiation were to exist, one could still picture the radioactive substances transforming it without themselves being altered. So far no experiment has confirmed or invalidated these hypotheses. (Becquerel, 1903, p. 15)

On the same occasion, Pierre Curie discussed the existing explanations of radioactivity. He presented a description of the earlier views of the Curies that is at variance with existing evidence:

> Since the beginning of our researchers we have noticed, Mrs. Curie and I, that to explain the phenomena it is possible to frame two distinct very general hypotheses that were presented by Mrs. Curie in 1899 and 1900. (Curie, 1903, p. 5)

The two hypotheses are then presented by Curie: the penetrating radiation hypothesis and the hypothesis of atomic disintegration. As has been shown above, the only hypothesis described in Marie Curie's early research papers is the first one. The second hypothesis does appear, *among several others* (for instance, a violation of the second law of thermodynamics) in the papers published in 1899 and 1900 by Marie Curie; but his only occurred after the penetrating radiation hypothesis had been challenged by the experiment of Elster and Geitel[6]. Now, in 1903, Pierre Curie seemed convinced that the atomic transformation hypothesis was the best explanation; and so he was careful enough to *conceal* that their initial assumption was the penetrating radiation hypothesis.

10. CONCLUSIONS

The penetrating radiation hypothesis had been very fruitful, in 1898, since it provided an explanation for the atomic property hypothesis that guided the discovery of polonium and radium. When the hypothesis encountered strong difficulties – such as Elster and Geitel's negative experiment in the late 1898 – the Curies maintained their hypothesis. When the conjecture was threatened by the discovery of the nature of the β radiation, in 1899, Pierre Curie and Georges Sagnac were able to sustain the hypothesis by showing that the secondary rays also contained particles with negative charge.

However, it is likely that this loyalty to the old hypothesis acted as a barrier to the understanding of radioactivity, in the next years. The Curies still kept their faith in this hypothesis at the time when Rutherford and Soddy began to develop the disintegration theory of radioactivity. They resisted the new theory, not because of their aversion to concrete, material hypotheses (as has been claimed) but because the new theory was incompatible with their own cherished explanation of radioactivity. In a few years, nonetheless, they had to give up their explanation because only Rutherford's theory of atomic disintegration and change could account for the wealth of evidence amassed by himself, by Frederick Soddy and by several other researchers.

---

[6] In her 1899 paper, Marie Curie described *five* (not two) groups of hypotheses for explaining the emission of energy by radioactive bodies (Sklodowska-Curie, 1899).

Although the traditional accounts of the work of the Curies do not emphasise their use of conjectures (see Weill, 1970; Wyart, 1970), I claim that their radioactivity researches were guided by some definite hypotheses, in the same way as Becquerel's research. In both cases, their scientific papers convey the feeling that their research was purely empirical and that they avoided any specific hypothesis, but that was not the case. Rutherford's hypotheses were perhaps more detailed and they were explicitly presented by him, in his paper. But that is just a difference of degree, not a qualitative difference between the attitudes of Rutherford and the Curies.

ACKNOWLEDGEMENTS

This paper was presented at the British Society for the History of Science Annual Conference University of Leeds, 15-17 July 2005. The author is grateful to the Brazilian National Council for Scientific and Technological Development (CNPq) and to the São Paulo State Research Foundation (FAPESP) for supporting this research; and to the *Fundo de Apoio ao Ensino, à Pesquisa e à Extensão* (FAEPEX) of the State University of Campinas, which provided support for participating in this Conference.

BIBLIOGRAPHIC REFERENCES

[ANONYMOUS]. Société Française de Physique. Séance du 3 Mai 1901. *Revue Générale des Sciences* **12**: 498-499, 1901.

BADASH, Lawrence. Radioactivity before the Curies. *American Journal of Physics* **33**: 128-135, 1965.

BECQUEREL, Henri. Sur les radiations émises par phosphorescence. *Comptes Rendus Hebdomadaires des Séances de l'Académie des Sciences de Paris* **122**: 420-421, 1896 (a).

BECQUEREL, Henri. Sur les radiations invisibles émises par les corps phosphorescents. *Comptes Rendus*

Hebdomadaires des Séances de l'Académie des Sciences de Paris **122**: 501-503, 1896 (b).

BECQUEREL, Henri. Sur quelquer propriétés nouvelles des radiations invisibles émises par divers corps phosphorescents. *Comptes Rendus Hebdomadaires des Séances de l'Académie des Sciences de Paris* **122**: 559-564, 1896 (c).

BECQUEREL, Henri. Sur une propriété nouvelle de la matière, la radio-activité. *Les Prix Nobel* **3**: 1-15, 1903.

CROOKES, William. *La genèse des éléments*. Translated by Gustave Richard. Paris: Gauthier-Villars, 1887.

CURIE, Ève. *Madame Curie*. Paris, Gallimard, 1939.

CURIE, Pierre. L'état actuel des recherches sur les substances radioactives. *Archive des Sciences Physiques et Natureles* [4] **10**: 388-389, 1900.

CURIE, Pierre. Conférence Nobel faite a Stockholm devant l'Académie des Sciences. *Les Prix Nobel* **3**: 1-7, 1903.

CURIE, Pierre & LABORDE, A. Sur la chaleur dégagée spontanément par les sels de radium. *Comptes Rendus de l'Académie des Sciences de Paris* **136**: 673-675,1903.

CURIE, Pierre & SAGNAC, Georges. Électrisation négative des rayons secondaires produits au moyen des rayons Röntgen. *Comptes Rendus Hebdomadaires des Séances de l'Académie des Sciences de Paris* **130**: 1013-1016, 1900.

CURIE, Pierre & SAGNAC, Georges. Électrisation négative des rayons secondaires issus de la transformation des rayons X. *Journal de Physique* [series 4] **1**: 13-21, 1902.

CURIE, Pierre & SKLODOWSKA-CURIE, Marie. Sur une substance nouvelle radioactive, contenue dans la pechblende. *Comptes Rendus de l'Académie des Sciences de Paris* **127**: 175-178, 1898.

CURIE, Pierre & SKLODOWSKA-CURIE, Marie. Sur la radioactivité provoquée par les rayons de Becquerel. *Comptes Rendus de l'Académie des Sciences de Paris* **129**: 714-716, 1899.

CURIE, Pierre & SKLODOWSKA-CURIE, Marie. Sur la charge électrique des rayons déviables du radium. *Comptes Rendus Hebdomadaires des Séances de l'Académie des Sciences de Paris* **130**: 647-650, 1900 (a).

CURIE, Pierre & SKLODOWSKA-CURIE, Marie. Les nouvelles substances radioactives et les rayons qu'elles émettent. Vol. 3, pp. 79-113, *in*: GUILLAUME, Charles-Édouard & POINCARÉ, Lucien (eds.). *Rapports Présentés au Congrès International de Physique réuni a Paris en 1900.* 3 vols. Paris: Gauthier-Villars, 1900 (b).

CURIE, Pierre, SKLODOWSKA-CURIE, Marie & BÉMONT, G. Sur une nouvelle substance fortement radioactive, contenue dans la pechblende. *Comptes Rendus de l'Académie des Sciences de Paris* **127**: 1215-1217, 1898.

DAVIS, J. L. The research school of Marie Curie in the Paris Faculty, 1907-14. *Annals of Science* **52**: 321-355, 1995.

ELSTER, Johann & GEITEL, Hans. Versuche an Becquerelstrahlen. *Annalen der Physik und Chemie* [2] **66**: 735-740, 1898.

JAUNCEY, G. E. M. The early years of radioactivity. *American Journal of Physics* **14**: 226-241, 1946.

JOLIOT-CURIE, Irène (ed.). *Oeuvres de Marie Sklodowska Curie / Prace Marii Sklodowskiej-Curie.* Varsovie / Warsawa: Panstwowe Wydawnictwo Naukowe, 1954.

JOLIOT-CURIE, Irène. Les carnets de laboratoire de la découverte du polonium et du radium. Pp. 103-124, *in*: SKLODOWSKA-CURIE, Marie. *Pierre Curie.* Paris: Denoël, 1955.

MALLEY, Marjorie. The discovery of atomic transmutation: scientific styles and philosophies in France and Britain. *Isis* **70**: 213-223, 1979.

MARTINS, Roberto de Andrade. Becquerel and the choice of uranium compounds. *Archive for History of Exact Sciences* **51** (1): 67-81, 1997.

NYE, Mary Jo. National styles? French and English chemistry in the nineteenth and early twentieth centuries. *Osiris* **8**: 30-49, 1993.

PERRIN, Jean. Rayons cathodiques et rayons de Röntgen. Étude expérimentale. *Annales de Chimie et de Physique,* [série 7], **11**: 496-554, 1897.[7]

RUTHERFORD, Ernest. Uranium radiation and the electrical conduction produced by it. *Philosophical Magazine* **47**: 109-163, 1899.

SAGNAC, Georges. Les expériences de M. H. Becquerel sur les radiations invisibles émises par les corps phosphorescents et par les sels d'uranium. *Journal de Physique Théorique et Appliquée* [3] **5**: 193-202, 1896.

SAGNAC, Georges. Luminescence et rayons X. *Révue Générale des Sciences* **9**: 314-320, 1898.

SAGNAC, Georges. Rayons X et décharge: généralisation de la notion de rayons cathodiques. *Comptes Rendus Hebdomadaires des Séances de l'Académie des Sciences de Paris* **130**: 320-323, 1900.

SAGNAC, Georges. Propagation des rayons X de Röntgen. *Annales de Chimie et de Physique* [série 7] **22**: 394-432, 1901 (a).

SAGNAC, Georges. Rayons secondaires dérivés des rayons de Röntgen. *Annales de Chimie et de Physique* [série 7] **22**: 493-563, 1901 (b).

SAGNAC, Georges. Relation des rayons X et de leurs rayons secondaires avec la matière et l'électricité. *Annales de Chimie et de Physique* [série 7] **23**: 145-198, 1901 (c).

SKLODOWSKA-CURIE, Marie. Rayons émis par les composés de l'uranium et du thorium. *Comptes Rendus Hebdomadaires des Séances de l'Académie des Sciences de Paris* **126**: 1101-1103, 1898 (a).

SKLODOWSKA-CURIE, Marie. Poszukiwania nowego metalu w pechblendzie. *Swiatlo* **1**, 1898 (b). Reproduced in

---

[7] This paper reproduces Perrin's PhD thesis.

pp. 49-56, JOLIOT-CURIE, *Oeuvres de Marie Sklodowska Curie*.

SKLODOWSKA-CURIE, Marie. Les rayons de Becquerel et le polonium. *Révue Générale des Sciences* **10**: 41-50, 1899.

SKLODOWSKA-CURIE, Marie. Sur le poids atomique du baryum radifère. *Comptes Rendus de l'Académie des Sciences de Paris* **131**: 382-384, 1900.

SKLODOWSKA-CURIE, Marie. Sur le poids atomique du radium. *Comptes Rendus de l'Académie des Sciences de Paris* **135**: 161-163, 1902.

SKLODOWSKA-CURIE, Marie. *Recherches sur les substances radioactives*. Paris: Gauthier-Villars, 1903.[8]

SKLODOWSKA-CURIE, Marie. *Pierre Curie*. Trad. Charlotte and Vernon Kellogg. With an introduction by Mrs. William Brown Meloney and autobiographical notes by Marie Curie. New York: MacMillan, 1923.

SODDY, Frederick. Radioactivity. *Annual Reports on the Progress of Chemistry for 1904 Issued by the Chemical Society* **1**: 244-280, 1905.

WEILL, Adrienne R. Curie, Marie (Maria Sklodowska). Vol. 3, pp. 497-503, *in*: GILLIESPIE, Charles Coulston (ed.). *Dictionary of Scientific Biography*. 16 vols. New York: Charles Scribner's Sons, 1970.

WYART, Jean. Curie, Pierre. Vol. 3, pp. 503-508, *in*: GILLIESPIE, Charles Coulston (ed.). *Dictionary of Scientific Biography*. 16 vols. New York: Charles Scribner's Sons, 1970.

---

[8] This was Marie Curie's PhD thesis.